# Approaches for Strengthening Total Force Culture and Facilitating Cross-Component Integration in the U.S. Military

AGNES GEREBEN SCHAEFER, JOHN D. WINKLER,
KIMBERLY JACKSON, DANIEL IBARRA,
DARRELL D. JONES, GEOFFREY MCGOVERN

Prepared for the Office of the Secretary of Defense
Approved for public release; distribution unlimited

NATIONAL DEFENSE RESEARCH INSTITUTE

For more information on this publication, visit www.rand.org/t/RR2143

**Library of Congress Cataloging-in-Publication Data** is available for this publication.
ISBN: 978-1-9774-0008-6

Published by the RAND Corporation, Santa Monica, Calif.
© Copyright 2020 RAND Corporation
**RAND**® is a registered trademark.

Cover: U.S. Air Force photo by Senior Airman Tristan D. Viglianco.

### Support RAND
Make a tax-deductible charitable contribution at
www.rand.org/giving/contribute

www.rand.org

# Preface

While all U.S. military services have strived to achieve greater total force integration and a stronger total force culture across their active and reserve components, significant impediments limit the achievement of these objectives. Since total force integration remains elusive, the issue continues to capture the attention of policymakers, who seek ways to overcome impediments and facilitate greater integration. This research identifies policies and practices that contribute to improved integration and a stronger total force culture, focusing on innovative approaches that could impart greater cross-component knowledge and awareness, and provide greater rewards for service members to work across components.

This report should be of interest to those concerned with active and reserve component organizational structure and integration. This research was sponsored by the Office of Reserve Integration within the Office of the Under Secretary of Defense for Personnel and Readiness and was conducted within the Forces and Resources Policy Center of the RAND National Defense Research Institute, a federally funded research and development center sponsored by the Office of the Secretary of Defense, the Joint Staff, the Unified Combatant Commands, the Navy, the Marine Corps, the defense agencies, and the defense Intelligence Community.

For more information on the RAND Forces and Resources Policy Center, see www.rand.org/nsrd/ndri/centers/frp or contact the Center director (contact information is provided on the webpage).

Comments or questions on this project report should be addressed to the project leaders, Agnes Schaefer, at schaefer@rand.org or 412-683-2300, extension 4488, and John Winkler, at jwinkler@rand.org or 703-413-1100, extension 5511.

# Contents

# Figures

# Summary

In an uncertain strategic and budgetary environment, cross-component integration remains a priority for defense policymakers as a continuing means for enhancing flexibility in capabilities, readiness, and force structure. Department of Defense (DoD) Directive 1200.17, *Managing the Reserve Components as an Operational Force*, outlines DoD's principles and overarching support for the implementation of policies supporting active component (AC) and reserve component (RC) integration. It specifically directs the service secretaries to "integrate AC and RC organizations to the greatest extent practicable, including the use of cross-component assignments, both AC to RC and RC to AC."[1]

This priority has been addressed most recently by national commissions addressing the future of both the Army and the Air Force.[2] While each of these sets of proposals provides ideas for enhancing integration and providing a greater total force culture, these specific proposals are neither complete nor fully reflective of all potentially relevant policies and practices. Further, these particular policy prescriptions are service specific and do not reflect broader insights that cut across services. Last, none of these efforts clearly define the desired purpose and end state for integration against which integration initiatives can be

---

[1]  DoD Directive 1200.17, *Managing the Reserve Components as an Operational Force*, October 29, 2008, p. 6.

[2]  National Commission on the Future of the Army (NCFA), *Report to the President and the Congress of the United States*, Arlington, Va., January 28, 2016; National Commission on the Structure of the Air Force (NCSAF), *Report to the President and Congress of the United States*, Arlington, Va., January 30, 2014.

evaluated. For these reasons, a more comprehensive analysis is needed of policies and practices that can contribute to the ultimate objective of improving total force integration and achieving a total force culture.

## Study Objective and Approach

The objective of this study is to provide insights on policies that can foster cross-component integration and incentives for cross-component service that contribute to the most effective total force possible, and benefit individual service members, as well as both the active and reserve components. The focus of this report is on factors that can increase cross-component knowledge and awareness, which contribute to achieving the larger goal of cross-component integration.

The study team approached this issue using a qualitative methodology consisting of focused literature reviews and facilitated discussions with senior service leaders and personnel managers from other U.S. government agencies. Our literature review covered social science research literature on organizational change; statutes and policies that govern personnel management and are relevant to accomplishing integration across components; how Goldwater-Nichols legislation attempted to foster "jointness" across the military services; and integration experiences in other U.S. government agencies, the private sector, and foreign militaries. Our discussions with DoD personnel and personnel managers in other U.S. government agencies sought to identify efforts within those organizations to promote integration across their different components, as well as the personnel management strategies used to implement those efforts. We then synthesized findings from the document reviews and informational discussions to identify potential strategies for DoD and the military services to consider for improving personnel integration across components. Not all findings were selected to be incorporated into our recommended potential actions; rather, we chose to focus on those findings that were particularly dominant in the literature, as well as findings that were particularly novel or significant in our case studies and discussions.

## Insights from U.S. Military Integration Efforts

In reviewing cross-component efforts already under way in the military services, we found common approaches used to foster cross-component integration. In one approach, AC and RC integration occurs at the organizational level, whereby elements of one component work with elements of a different component to perform, or train to perform, an operational mission. Examples include multicomponent units (MCUs) (such as those found in the Army and Coast Guard)[3] and associate units (such as those found in the Air Force and the Army).[4]

In a second approach, AC and RC integration occurs at the individual level, whereby service members from one component assume positions in another component as part of a unit or a headquarters staff. Each service's Individual Mobilization Augmentee (IMA) program and the Marine Corps' Inspector and Instructor (I&I) program (in which AC members are embedded in RC units to oversee RC readiness) provide examples of individual integration across the active and reserve components. Other examples include embedding RC members in a single unit under an AC chain of command, and cross-component command positions (such as those found in the Air Force, Marine Corps, and Coast Guard). Across the services, efforts to promote integration at the individual level typically focus on three elements of personnel management: assignments, promotion, and pay and benefits. The services all have unique challenges when it comes to incentivizing assignments to foster greater AC and RC integration.

### Lessons Learned from Integration Efforts in the Services

Looking across these efforts, we identified factors that are commonly perceived as facilitating and inhibiting efforts to enhance cross-component integration. Integration efforts are facilitated when

- initiatives consider unique service force structures and RC competencies

---

[3] MCUs are units composed of members from at least two components.

[4] Associate units are reserve and active units that integrate and often share equipment.

- RC capabilities are included in service strategic planning
- leadership sets the tone, message, and pace regarding integration efforts
- initiatives define an end state and are implemented deliberately
- incentives are used to attract individuals to cross-component assignments (e.g., command opportunities, geographic location, and financial incentives).

Circumstances that inhibit cross-component integration include

- cultural differences across the components
- statutory and funding constraints
- lack of recognition or reward for serving in cross-component assignments
- prescriptive and rigid career development paths that inhibit cross-component talent management strategies
- lack of formal evaluation, limiting the ability to demonstrate benefits
- failures or errors in implementation.

## Joint Integration: A Potential Analogue for Cross-Component Integration

As part of our review, we also examined one of the major DoD reorganization efforts of the last century: the Goldwater-Nichols Department of Defense Reorganization Act of 1986—congressional legislation aimed at ensuring greater cooperation and integration across the military services. Our purpose for this analysis was to determine whether the mechanisms employed to foster "jointness" are applicable for facilitating cross-component integration. One facet of this legislation that is relevant to cross-service integration is the personnel management practices used to induce and entice officers to engage in more joint operations planning and management, thereby fostering a cultural change that would embrace jointness. Lessons derived from DoD's experience with joint integration may offer guideposts to future DoD cross-component integration efforts. These lessons include the following:

1. Cultural change was realized. Although Congress provided the initial impetus through statutory direction, top leadership in DoD communicated the importance of and provided visible direction for these changes.
2. Changes to assignment requirements ensured that the services sent their best individuals to joint positions.
3. Changes to promotion requirements, requiring joint positions for promotion to senior rank, ensured that the best individuals applied for joint positions.
4. Changes to training and education also ensured jointness across services. By further emphasizing joint training and education across the services, individuals became increasingly familiar with the capabilities, cultures, and processes in the other services— fostering a better understanding of a broader DoD culture.
5. Incentives at both the service level and the individual level were automatically incorporated into the changes to the assignment, promotion, and training and education requirements ushered in by Goldwater-Nichols.
6. Assignment, promotion, and training and education requirements worked together with incentives as a system to promote organizational objectives and to create an environment in which serving in joint assignments was viewed as a necessary and desirable part of an officer's career development.

## Insights from Other Integration Efforts

We also analyzed the integration and rotational programs implemented by civilian U.S. government agencies and the private sector, as well as the integration experiences of foreign militaries. Although the cultures and missions of these other organizations vary and are not exact analogues to the U.S. military, their experiences offer useful insights into ways to achieve greater integration across large, and sometimes disparate, organizations.

**Figure S.1**
**Approaches Found in Other Organizations' Integration Efforts**

| Civilian Agencies | Private Sector | Foreign Militaries |
|---|---|---|
| • Link cross-component assignments to future assignments<br>• Give points for promotion for cross-component assignments<br>• Provide an additional retirement annuity benefit for cross-component assignments<br>• Offer early bidding for next assignments after cross-component assignments<br>• Make additional education and training opportunities available after cross-component assignments<br>• Provide choice in duty location after cross-component assignments<br>• Mandate cross-component assignments for promotion | • Use rotational assignments to expand individuals' knowledge of broader enterprise operations<br>• Require new hires to rotate around the organization before deciding which component to work in<br>• Offer midcareer gap years or sabbaticals to broaden their careers | • Increase permeability across components to harness skills and expertise<br>• Change conditions of service to allow for more flexibility across components<br>• Make training requirements the same across AC and RC to facilitate utilization of personnel across components<br>• Train on the same equipment, regardless of component |

From this analysis, we identified approaches to integration that could potentially be applied in the U.S. military to further integrate the active and reserve components (see Figure S.1).

These approaches aim to foster integration by offering opportunities for career-broadening experiences, standardizing training across organizations, and offering more flexibility to employees in their career progression paths. These types of approaches align with the direction in which some of the services are already headed. While current personnel policies in some of the services are too rigid to accommodate some of the approaches above, the current trends toward permeability and individual talent management open the door to consider more flexible personnel management approaches to AC/RC integration, including those cited in the report or variations.

## Crosscutting Lessons Learned from These Integration Efforts

When looking across the experiences of U.S. civilian agencies, the private sector, and foreign militaries, several crosscutting lessons can be drawn:

- Fostering integration requires a shift in culture and leadership buy-in.
- Required rotations can improve retention and facilitate a holistic understanding of enterprise operations.
- Integration is often easier with more junior employees.
- Financial incentives are not always the most compelling.

## Applying Findings to Better Facilitate AC/RC Integration

In thinking about how to apply these findings to identify ways in which the services can better facilitate AC/RC integration, we developed a multifaceted strategic human resources framework composed of elements that can facilitate integration and achieve organizational goals. This framework includes (1) changes to personnel management policies related to assignment, (2) changes to personnel management policies related to promotion, (3) changes to training and education, and (4) use of incentives to make cross-component service more attractive and rewarding to service members. It also includes broader structural and statutory issues that need to be addressed as part of such strategies. Within each of these categories, we identified numerous potential actions that could facilitate deeper AC/RC integration. These potential actions are summarized in Figure S.2.

- **Changes to assignment processes.** One of the first steps that the services could take is to identify and expand the number of positions that are suitable for cross-component assignments. The Goldwater-Nichols Act established the number of officers who would serve in joint duty assignments, and it created a system of education and experience requirements that are prerequisite to service as officers in joint specialties. A similar set of requirements

**Figure S.2**
**Summary of Potential Actions to Increase AC/RC Integration**

**Changes to Assignment Processes**

- Identify and expand positions suitable for cross-component assignments
- Improve screening for cross-component assignments to ensure high-quality candidates
- Utilize potential incentives as needed to fill cross-component assignments
- Consider and mitigate effects of changes on AC and RC career paths

**Changes to Training and Education**

- Expand opportunities for cross-component training

**Changes to Promotion Processes**

- Clarify precepts and board changes
- Expand board membership
- Develop AC/RC qualification system

**Incentives for Cross-Component Assignments**

- Identify appropriate combination of monetary and nonmonetary incentives
  Monetary incentives include:
  o assignment and incentive pays
  o subsidies for housing
  o subsidies for childcare and family benefits
  Nonmonetary incentives include:
  o cross-component assignments linked to future assignments
  o work-life balance
  o award ribbon or qualification for cross-component assignments

**Structural Changes**

- Ensure Total Force strategy guides integration efforts
- Implement talent management workforce strategy
- Address legal/regulatory challenges and undertake efforts to develop mitigation strategies
  o Duty status reform
  o Improve scrolling process (the transfer of a service member from one component to another)
- System changes (e.g., fully implement Integrated Pay and Personnel Systems)

could potentially be established for cross-component assignments. However, it is important to note that during our discussions with senior leaders and other representatives of the services, many expressed hesitations to create additional "cross-component" requirements since it is already difficult for service members to meet all of their career requirements in order to advance.

- **Changes to promotion processes.** Like any incentive, changes to promotion processes could drive service member behavior regarding AC/RC integration. An excellent example of incentives driving individual behavior in the personnel management arena resulted from the enactment of Goldwater-Nichols. When DoD leadership began tracking promotion rates for joint officers, when these officers had to be promoted at specific promotion rates, and when joint qualification became a requirement for promotion to general and flag officer rank, officers began to pursue joint assignments. However, impacting the promotion process directly should not be taken lightly.
- **Changes to training and education.** By further emphasizing cross-component training and education, individuals will become increasingly familiar with the capabilities housed in the other components, as well as different cultures and processes across the components. This could not only facilitate better interoperability across services but also foster a better understanding of a broader DoD culture.
- **Incentives.** Both civilian and military organizations have used monetary and nonmonetary incentives to change individuals' preferences for assignments. Monetary incentives include assignment and incentive pays, subsidies for housing, and subsidies for childcare and family benefits. Nonmonetary incentives include cross-component assignments linked to future assignments, work-life balance (e.g., accommodate preferred assignment locations; flexible schedules, comp time, vacations; facilitate seamless transition across components; expand and publicize Career Intermission Program [CIP]); and award ribbon, or qualification for cross-component assignments.

- **Structural changes.** In addition to the changes in the previous four facets of our framework, strengthening AC/RC integration will also require potential foundational structural changes. These include potential changes to DoD strategy and doctrine, systems, and processes, as well as legal and regulatory changes and force structure changes.

## Conclusions: What the Services Should Do Next

Our findings indicate that it is possible to enhance cross-component knowledge and awareness and further develop a total force culture. However, additional steps need to be taken to modify and align personnel policies to achieve these objectives. Most importantly, changes to assignment and promotion policies are critical for accomplishing these objectives. Our findings also indicate that various incentives (monetary and nonmonetary) can facilitate cross-component integration.

Senior DoD and service leadership can take several steps to help facilitate deeper AC/RC integration. First, they should decide whether integration is a priority. If so, top service leadership should clarify the purpose of integration, define the ultimate end state, and establish goals and benchmarks for furthering integration. Next, the services should undertake a review of current assignment and promotion policies and determine how these will change. This includes (1) identifying positions for cross-component assignments, (2) determining the number and grade levels of assignees, (3) implementing a program of incentives as needed, and (4) altering promotion policies and practices to reward cross-component service. Last, DoD should continue to seek structural changes that overcome legal and regulatory barriers to integration while implementing system changes that could facilitate AC/RC integration. In addition, the services and DoD should facilitate further integration efforts by defining the purpose of AC/RC integration efforts, fostering a shift in culture and leadership buy-in, tailoring integration efforts to unique service force structures and RC competencies, and evaluating integration initiatives.

# Acknowledgments

This report would not have been possible without the assistance of scores of people. We thank Mr. Matthew Dubois, the former Deputy Assistant Secretary of Defense (Reserve Integration), for sponsoring our study. We also thank our project monitors, Col Ernest Ackiss and CAPT Eric Johnson, who worked closely with us over the course of the study and enabled us to gain access to important sources of information, documentation, and people.

We would also like to thank those who gave their time to discuss this topic and share their ideas with our research team. This includes senior leaders and representatives from the services, as well as civilian organizations including the Defense Intelligence Agency, Department of State, and the Office of the Under Secretary of Defense (Policy).

We also benefited from the contributions of many RAND colleagues, including Greg Schumacher, Al Robbert, Craig Bond, and Lisa Harrington, who provided helpful feedback in the form of formal peer reviews. We also thank Barbara Bicksler, who assisted in the preparation of this report. We thank them all but retain full responsibility for the objectivity, accuracy, and analytic integrity of the work presented here.

# CHAPTER ONE

# Introduction

## Background and Study Purpose

From the time that Sec. Melvin Laird first established the concept of the total force in 1970, Department of Defense (DoD) policies have sought greater integration between active and reserve forces. A fuller integrated force of active component (AC) and reserve component (RC) personnel is seen as beneficial from the perspective of force economy and effectiveness, under the presumption that a mixture of AC and RC capabilities and personnel can be optimized to achieve maximum effective capacity at the lowest cost. Integration is also viewed favorably as a means for developing well-rounded leaders with a stronger awareness and understanding of the full capabilities of their military service, thereby furthering a total force culture. Finally, integration is seen as a necessary consequence of active duty force reductions and a greater continuing reliance on reserve forces to perform operational missions.[1]

Integration can occur at multiple levels, including the unit level, whereby elements of one component are embedded within elements of another component. Integration can also occur at the individual level, whereby members from one component serve alongside members of a different component. These cross-component assignments can include staff and line positions. For example, the service member may

---

[1] NCFA, NCFA Operation Subcommittee Report, Open Meeting, *The Total Force Policy and Integration of Active and Reserve Units (Multiple Component Units-MCU)*, Arlington, Va., December 17, 2015.

be assigned to a headquarters or to a command group or can fill a vacant position within an operational unit.

Although integration has been viewed as beneficial for achieving Total Force policy objectives, and while efforts to more fully integrate have been under way in each of the military services for some time, barriers and constraints to integration continue to exist. Foremost among these barriers are those rooted in cultural aspects of each service component and established within each component's history, identity, and experiences. These cultural differences may be further amplified by attitudes and perceptions of personnel in other components to include, for example, perceptions of competence, trustworthiness, and capacity to perform. Some prominent examples of cross-component conflict and mutual distrust over the past 25 years include the aftermath of decisions to not deploy activated Army National Guard (ARNG) combat units during the first Persian Gulf War[2] and more recent conflicts within the Army and the Air Force following attempts by military and civilian leadership to change personnel end strengths, adjust force structure, or reassign missions from the then-current status quo.[3]

In addition to cultural barriers to integration, structural impediments also act to limit the ability of policymakers to achieve cross-component integration of personnel. These include statutory restrictions that constrain movement of personnel across components (Title 10), laws related to the functions and purposes of the National Guard (Title 32), and service-specific policies and practices related to personnel selection, assignment, and promotion—many of which may discourage integration across components. Any future efforts to further integrate the active and reserve components will need to address these cultural and statutory barriers.

In an uncertain strategic and budgetary environment, cross-component integration remains a priority for defense policymakers as a

---

[2]   Robert L. Goldrich, *The Army's Roundout Concept After the Persian Gulf War*, Congressional Research Service Report for Congress, October 22, 1991.

[3]   Loren Thompson, "Shrinking Army Fights National Guard for Vital Combat Helicopters," *Forbes*, June 30, 2014; Christian Davenport, "Air Force Plan to Get Rid of A-10s Runs into Opposition," *Washington Post*, April 10, 2014.

continuing means for enhancing flexibility in capabilities, readiness, and force structure. Recently, national commissions addressing the future of the Army and of the Air Force have reinforced this priority.[4] Each of these commissions has proposed to increase integration through changes in personnel policy. For example, the National Commission on the Future of the Army (NCFA) recommended that the Secretary of the Army should review and assess officer and noncommissioned officer (NCO) positions from all components for potential designation as integrated positions. In an attempt to foster an Army total force culture and expand knowledge about other components, individuals from all components could fill these integrated positions.[5] The commission also recommended that the Secretary of the Army should develop selection and promotion policies that incentivize Regular Army, ARNG, and Army Reserve (USAR) assignments across components and within multicomponent units (MCUs).[6]

The National Commission on the Structure of the Air Force (NCSAF) recommended that the Air Force "should develop and supervise implementation of a pilot project for a Continuum of Service."[7] The "continuum of service" personnel management construct calls for a more "seamless" flow between full- and part-time service and more opportunities for individuals to move between positions in the AC and RC.[8] Similarly, the concept of "permeability," which is advanced in the Army commission report and further amplified in recent DoD "Force of the Future" proposals,[9] envisions mechanisms by which service members

---

[4]  NCFA, 2016; NCSAF, 2014.

[5]  See Recommendation 27, NCFA, 2016, p. 65.

[6]  See Recommendation 28, NCFA, 2016, p. 65.

[7]  NCSAF, 2014, p. 51.

[8]  John D. Winkler et al., "A 'Continuum of Service' for the All-Volunteer Force," in Barbara A. Bicksler, Curtis L. Gilroy, and John T. Warner, eds., *The All-Volunteer Force: Thirty Years of Service*, Washington, D.C.: Brassey's, Inc., 2004, p. 300.

[9]  Force of the Future initiatives were introduced by former Secretary of Defense Ashton Carter and were designed to increase DoD's permeability to new people and ideas. See Ashton Carter, Secretary of Defense, "Force of the Future: Maintaining Our Competitive Advantage in Human Capital," memorandum for secretaries of the military departments,

may easily transfer between components. Integrating the active and reserve components is not a new issue to the services and DoD. However, incentivizing AC personnel to serve within the RC, along with decreasing the administrative burden on service members desiring to transition between the components, has recently garnered much interest from policymakers and senior leaders.

While both the Army and Air Force commissions' proposals offer ideas for enhancing integration and providing a greater total force culture, these specific ideas are neither complete nor fully reflective of all potentially relevant policies and practices. Further, these particular policy prescriptions are service specific and do not reflect broader insights that cut across services. For these reasons, a more comprehensive analysis is needed of policies and practices that can contribute to the ultimate objective of improving total force integration and achieving a total force culture.

## Study Objective and Approach

The objective of this study is to provide insights on policies that can foster cross-component integration and incentives for cross-component service that contribute to the most effective total force possible, and benefit individual service members, as well as both the active and reserve components. The focus of this report is on factors that can increase cross-component knowledge and awareness, which contribute to achieving the larger goal of cross-component integration. This project takes as a starting point the recommendations from the recent NCFA and NCSAF reports to improve AC/RC cross-component integration, and then provides specific suggestions for achieving greater integration through new assignment and promotion policies, as well as potential structural changes (including legal, regulatory, and information systems changes).

---

November 18, 2015. Also see U.S. Department of Defense, "Force of the Future: Whatever You Want to Do, You Can Do in Service to Your Country"; Cheryl Pellerin, "Carter Unveils Next Wave of Force of the Future Initiatives," *DoD News*, June 9, 2016.

The study team approached this issue using a qualitative methodology consisting of focused literature reviews and facilitated discussions with senior service leaders and personnel managers. The literature review included (1) research on organizational change that identifies factors that facilitate or inhibit integration within an organization, and (2) research on the factors associated with successful mergers and consolidations, focusing particularly on personnel practices that help make the merger or consolidation successful.

As part of our review of written materials, we looked at statutes and policies that govern personnel management and are relevant to accomplishing integration across the active and reserve components, and we also examined documents and regulations describing previous attempts at fostering integration in the military services. We also undertook a review of how Goldwater-Nichols legislation attempted to foster "jointness" across the military services by assigning military officers to positions outside their home service and within joint organizations. We examined how these efforts contributed to development of greater knowledge and awareness across military services, and whether these experiences provide insights for how greater AC/RC cross-component integration might be achieved. In addition, we also reviewed the literature on the integration experiences in other U.S. government agencies, the private sector, and select foreign militaries.

The study approach also included informational discussions with personnel from the military services and other U.S. government agencies.[10] Discussions with senior DoD leaders focused on integration efforts in the specific military services, including programs and personnel management mechanisms that address AC/RC integration directly, as well as initiatives that are intended to provide service members with career-broadening experiences that expand their institutional knowledge and awareness. These discussions sought to (1) identify relevant programs within those organizations that seek to build a more cohesive and integrated civilian workforce and provide employees with

---

[10] This study received RAND Human Subject Protection Committee approval to proceed with informational discussions on September 2, 2016. We conducted discussions with 31 individuals during the course of this study.

enhanced knowledge and awareness of agency-wide missions and operations, and (2) understand what these programs seek to accomplish, how they are intended to work, and barriers and constraints they face in accomplishing program objectives.

We took notes during all of our information discussions and then coded this information to identify barriers to integration, approaches used by the military services and other government agencies to provide career-broadening experiences, and lessons learned (including implementation challenges). We then synthesized findings from the document reviews and informational discussions to identify potential actions that DoD and the military services could take to improve personnel integration across components. These include changes to personnel management policies related to assignment, promotion, and training and education, as well as incentives to make such service more attractive and rewarding to service members. They also include broader structural and statutory issues that need to be addressed as part of such strategies. Not all findings were selected to be incorporated into our recommended potential actions; rather, we chose to focus on those findings that were particularly dominant in the literature, as well as findings that were particularly novel or significant in our case studies and discussions.

## Organization of Report

The remainder of this report is organized into five chapters. Chapter Two presents the results of our review of U.S. military integration efforts. Chapter Three explores experiences in fostering "jointness" across services as an analogue for fostering integration across components. Chapter Four offers insights from the literature on organizational change and integration, as well as from other integration efforts within the U.S. government, the private sector, and foreign militaries. Chapter Five applies our findings to the development of specific policy and structural changes that could strengthen AC/RC integration and foster the development of a stronger total force culture. Chapter Six contains our conclusions and recommendations for next steps that the services and DoD can take to help facilitate deeper integration across active and reserve components.

# CHAPTER TWO

# Insights from U.S. Military Integration Efforts

Insights drawn from previous and current efforts to enhance cross-component integration in the military provide a starting point for identifying promising directions for future integration. Efforts to integrate the AC and RC are not new, and while the nature of AC/RC integration differs among the military services, their experiences provide evidence that greater operational integration of capabilities, better understanding of capabilities across components, and greater flexibility in utilizing personnel and skills across components can be achieved. However, these integration efforts have also fallen short in important ways, and they provide lessons learned for future integration efforts. This chapter begins by describing the legal context for military integration efforts and then describes recent and current military integration efforts. It describes how each of the military services (including the U.S. Coast Guard) has approached cross-component integration, and draws on these experiences to identify general approaches used, lessons learned, and barriers facing future integration efforts.

## Legal Context for Military Integration Efforts

The U.S. armed forces are governed by Title 10 of the United States Code (U.S.C.). Title 10 represents the permanent laws passed by Congress and signed by the President regarding the personnel, equipment, and operation of the U.S. military. It covers aspects of the armed forces ranging from staffing (including prescribing positions for the Secretary of Defense through the heads of the veterinary and dental corps of the services), to placing limits on the number of general and flag officers of various grades, to specifying medical and retirement benefit systems.

The statutory purpose of the RCs of the armed forces is explained in 10 U.S.C. 10102: "The purpose of each RC is to provide trained units and qualified persons available for active duty in the armed forces, in time of war or national emergency, and at such other times as the national security may require, to fill the needs of the armed forces whenever more units and persons are needed than are in the regular components."[1] Administration of the RCs is statutorily given to the secretaries of each service by 10 U.S.C. 10202.

Title 10, then, could hinder integration of the active and reserve forces if provisions of law place substantive or procedural hurdles in the path of integration efforts. These hurdles could include provisions in statutes that prohibit certain types of activities. For example, 10 U.S.C. 10213 states, "[N]o person may be a member of more than one RC at the same time."[2] They could also entail statutorily required processes, including but not limited to budgeting and appropriations processes; required promotion, training, and retirement management processes; and processes governing the call of reserve forces to active duty.

Given the broad scope of Title 10—covering all areas of DoD operations—it is likely that some provisions of law could be viewed as obstacles to integration efforts. While nothing in extant law specifically prohibits a *general* sense of AC/RC integration, specific efforts toward integration might run afoul of statutory restrictions. For example, 10 U.S.C. 12304(b) is a specific provision that restricts when reserve forces can be ordered into active duty for preplanned missions in support of the combatant commands. According to the law, reserve units can be ordered into such specified active duty only if the manpower and related costs for that duty are "specifically included and identified in the defense budget materials for the fiscal year or years in which such units are anticipated to be ordered to active duty." Furthermore, "not more than 60,000 of the RCs of the armed forces may be on

---

[1]   See 10 U.S.C. 10102, "Purpose of Reserve Components."

[2]   See 10 U.S.C. 10213, "Reserve Components: Dual Membership Prohibited."

active duty under this section at any one time."[3] While DoD has other authorities allowing the leadership to call up reserve forces to augment the AC, in this provision we see both absolute limits on manpower and budgetary restrictions that potentially constrain the degree to which cross-component integration efforts can take place under specified circumstances. This is not to say that such limitations are not wise; rather, this example demonstrates the type of specificity in Title 10 that could potentially affect integration efforts by the services.

In fact, in the 2016 NCFA report, 10 U.S.C. 12304(b) was highlighted as an area of law important to Army integration efforts. The report recognized that Sec. 12304(b) envisions potentially greater use of the reserve forces, but commented that the special budgetary processes resulted in less than full funding for 12304(b) force deployment: "The Total Force Policy must be resourced if it is going to be effective, and the absence of adequate 12304(b) funding will limit using ARNG and USAR forces on missions for which they are ideally suited."[4]

Whereas Title 10 governs the armed forces of the United States, Title 32 governs the National Guard (meaning the ARNG and the Air National Guard [ANG]). National Guard forces can operate as either state or federal forces, though they operate under the control of the states until and unless activated for federal service. Title 32 U.S.C.102 outlines the general policy regarding the National Guard:

> Whenever Congress determines that more units and organizations are needed for the national security than are in the regular components of the ground and air forces, the ARNG and the ANG, or such parts of them as are needed, together with such units of other reserve components as are necessary for a balanced force, shall be ordered to active federal duty and retained as long as so needed.[5]

---

3   See 10 U.S.C. 12304b, "Selected Reserve: Order to Active Duty for Preplanned Missions in Support of the Combatant Commands."

4   NCFA, 2016, p. 66.

5   See 32 U.S.C. 102, "General Policy."

The language is similar to that for the Title 10 RCs in 10 U.S.C. 10102, above. However, unlike Title 10, where active and reserve components are entirely federal entities, integration of National Guard forces, either individually or at the unit level, into federal missions can present unique federalism challenges. In addition, integrating Title 10 forces into National Guard units can also be challenging. Title 32 U.S.C. 315 authorizes the secretaries of the Army and Air Force to detail a Title 10 officer or enlisted member to a state National Guard. However, a Title 10 officer detailed to a state National Guard may accept a state commission in the ARNG or the ANG only with the consent of the President and the state governor involved.

Integration of National Guard forces is an inherent need given the federal and state dimensions of Title 32. Title 32 U.S.C. 315, for example, *requires* the secretaries of the Army and the Air Force to detail commissioned officers of the Regular Army and Regular Air Force to duty with the ARNG and ANG of each state, the commonwealth of Puerto Rico, the District of Columbia, Guam, and the Virgin Islands. The secretaries have discretion, however, whether they detail enlisted members to the states.[6] Title 32 U.S.C. 301 prescribes the conditions for federal recognition of enlisted members of a state National Guard unit;[7] Sec. 305 specifies federal recognition requirements for state commissioned officers.[8] In addition, Title 32 specifies conditions and processes for appropriation of resources, equipment, training, and courts-martial.

This dual nature of the National Guard as both federal and state forces presents a different set of challenges to integration than those of the Title 10 active and reserve forces. Title 10 U.S.C. 12301(d) hints at this dual nature by recognizing the need for gubernatorial consent to call National Guard members to active duty:

---

[6]  See 32 U.S.C. 315, "Detail of Regular Members of Army and Air Force to Duty with National Guard."

[7]  See 32 U.S.C. 301, "Federal Recognition of Enlisted Members."

[8]  See 32 U.S.C. 305, "Federal Recognition of Enlisted Members: Persons Eligible."

At any time, an authority designated by the Secretary concerned may order a member of a RC under his jurisdiction to active duty, or retain him on active duty, with the consent of that member. However, a member of the ARNG or ANG may not be ordered to active duty under this subsection without the consent of the governor or other appropriate authority of the state concerned.[9]

Furthermore, National Guard members frequently find themselves in a transitory status. For example, when ANG members are performing noncontingency alert duty, care needs to be taken to convert Title 32 status ANG members to Title 10 status. While these changes in duty status are regular and subject to well-documented provisions and processes, including documentation that members of the ANG performing alert duties have to sign a Title 10 consent statement prior to performing alert duty,[10] they represent the Title 10 and Title 32 hurdles that may make some integration efforts more challenging. However, it should also be noted that innovative solutions such as "automatically executing orders" (which are orders that automatically convert a service member from Title 10 status to Title 32 status, or vice versa, under specific circumstances) have been helpful in navigating some of the challenges associated with changes in duty statuses.

Title 10 and Title 32, in both absolute wording and interpretation, can limit the degree and extent of cross-component integration and associated policies. Senior DoD leaders need to decide what kind of integration is needed to field the most effective and affordable total force and then either decide how to achieve that within current law or request changes to applicable laws that inhibit the achievement of the desired end state.

---

[9] See 10 U.S.C. 12301, "Reserve Components Generally."

[10] See, for example, Air National Guard Instruction (ANGI) 36-101, *Air National Guard Active Guard/Reserve (AGR) Program*, June 3, 2010; ANGI 10-203, *Air National Guard Alert Resource Management*, February 22, 2012.

## Recent Military Cross-Component Integration Efforts

DoD, the Joint Staff, and the individual services have published recent doctrine, policy, and instructions that outline efforts to further integrate the active and reserve components. DoD Directive 1200.17, *Managing the Reserve Components as an Operational Force*, outlines DoD's principles and overarching support for the implementation of policies supporting AC and RC integration. Integration efforts outlined in this directive encourage both unit and individual integration of AC and RC personnel. It specifically directs the service secretaries to "integrate AC and RC organizations to the greatest extent practicable, including the use of cross-component assignments, both AC to RC and RC to AC. Such assignments should be considered as career enhancing and not detrimental for a service member's career progression."[11]

This directive also provides impetus for expanded service across components. It directs that both monetary and nonmonetary incentives should be utilized by the services to retain and promote service within the RC above the minimum participation level. Monetary incentives include bonuses, which can be used to retain recently separating AC personnel who desire to continue to serve in the RC. Nonmonetary incentives, such as access to affordable health insurance (TRICARE), desirable geographic assignments, and access to military base support functions (e.g., commissaries, base exchanges, child development centers, gyms), are benefits that positively affect a service member's quality of life.

We next describe how individual services have approached the objective of fostering further integration, and provide key takeaways from their integration initiatives.

### U.S. Air Force

Beginning with its efforts dating back to the 1960s to establish "associate units" (in which AC units and RC units share equipment), the Air Force has a history of strong cross-component initiatives. In 2005,

---

[11] DoD Directive 2008, p. 6.

the Air Force developed a plan for its total force structure, including a reorganization of the ANG, over the next 20 years.[12] Air Force Policy Memorandum (AFPM) 90-10, *Total Force Integration (2016)*, outlines recent Air Force perspectives on total force integration, including organizational roles and responsibilities. AFPM also indicates that

> [w]here it makes sense to do so, the Air Force must increase opportunities for component integration through enhanced cooperation in planning and programming, greater total force presence on staffs, Total Force Associations, and organizationally interchangeable positions to be filled by airmen of any component.[13]

As of early 2016, more than 78 total force integration proposals were being pursued by the Air Force, including 41 recommended by the NCSAF.[14] These included cross-component utilization of personnel, cross-flowing skilled officers between the components, implementing dual-status commanders, and integrating support staffs.[15] The Air Force's cross-component initiatives include approaches to cross-component integration at both the unit level and the individual level. We discuss several of these initiatives below.

### Approaches to Cross-Component Unit-Level Integration

The Air Force has developed five models of cross-component unit integration:

---

[12] U.S. Government Accountability Office, *Defense Management: Fully Developed Management Framework Needed to Guide Air Force Future Total Force Efforts*, GAO-06-232, Washington, D.C., January 31, 2006.

[13] U.S. Department of Air Force, Air Force Policy Memorandum (AFPM) 90-10, *Total Force Integration*, October 27, 2016. U.S. Department of Air Force, Air Force Instruction 90-1001, *Special Management: Planning Total Force Associations*, January 9, 2017, implements AFPM 90-10 for planning Total Force Associations and the development of Total Force Association proposals.

[14] Secretary of the Air Force Public Affairs, "Air Force Continues to Pursue Total Force Integration," *U.S. Air Force*, March 11, 2016.

[15] Secretary of the Air Force Public Affairs, 2016.

1. Classic Associate Unit, where an AC unit retains principal responsibility for a weapon system or systems, which it shares with one or more RC units. The AC and RC units retain separate organizational structures and chains of command.

2. Active Associate Unit, where an RC unit has principal responsibility for a weapon system or systems, which it shares with one or more AC units. RC and AC units retain separate organizational structures and chains of command.

3. Air Reserve Components Associate Unit, where two or more Air Reserve Component units integrate with one retaining principal responsibility for a weapon system or systems, which are shared by all. Each unit retains separate organizational structures and chains of commands.

4. Integrated Associate Unit, which is similar to the classic associate model; however, members of all components contribute to one unit mission with administrative control and support provided by the respective component via detachments.

5. Fully Integrated Unit, where members from different components make up a single organization, falling under the same chain of command.[16]

Classic associations are very common in mobility and training missions, less common in fighter missions, and also found in some nonflying missions, such as Rapid Engineer Deployable Heavy Operational Repair Squadron Engineers (RED HORSE) units and air or space operations centers.[17] For instance, the 219th RED HORSE Squadron of the Montana National Guard is associated with the 819th RED HORSE Squadron, an active duty squadron. Both squadrons are based at Malmstrom Air Force Base (AFB) in Montana. As 219th RED HORSE Squadron Commander Col. Rusty Vaira explained, "We're a classic association with the active duty 819th RED HORSE Squadron,

---

[16] See U.S. Department of Air Force, Air Force Policy Directive (AFPD) 90-10, *Total Force Integration Policy*, June 16, 2006 (certified current July 31, 2014).

[17] See Albert A. Robbert, James H. Bigelow, John E. Boon, Jr., Lisa M. Harrington, Michael McGee, S. Craig Moore, Daniel M. Norton, and William W. Taylor, *Suitability of Missions for the Air Force Reserve Components*, Santa Monica, Calif.: RAND Corporation, RR-429-AF, 2014, p. 43.

we share space, we share training, we share equipment as a total force initiative."[18]

The 173rd Fighter Wing at Kingsley Field ANG Base in Klamath Falls, Oregon, is an example of an active associate unit. The Oregon ANG owns the F-15Cs assigned to the 173rd Fight Wing that are used to train all of the Air Force's F-15C pilots—regardless of component. To aid the ANG in its training mission, an AC detachment of pilots, maintainers, and administrative personnel were assigned to Kingsley Field ANG Base. A senior Air Force leader specifically mentioned Kingsley Field in our discussion with him, and voiced caution about the second-order effects of active associations.

> There have been lots of growing pains associated with this model because Kingsley Field ANG Base does not have the same services that you would typically find at an AC installation, such as childcare centers, healthcare facilities, and gyms, and the housing rental market in Klamath Falls is extremely expensive.[19]

As a result, the Air Force chose to provide additional subsidies to cover these additional costs. This example illustrates that integrating AC and RC forces often has unforeseen challenges, and that senior leaders and policymakers must remain flexible and invested in these efforts to ensure that integration goals are achieved.

The Integrated Wing, or I-Wing, program is a pilot program proposed by the NCSAF that would implement a fully integrated cross-component unit. The NCSAF outlined the goals of this program by stating "associate units should have a single integrated chain of command. . . ."[20] This would be accomplished by having different component squadrons (such as an AC fighter squadron along with an RC fighter squadron) aligned under the same Operations Group. In February 2016, Air Force Sec. Deborah James announced that the first

---

[18] Eric Peterson, 120th Airlift Wing Public Affairs Office, "Restructuring Brings New Capabilities to the 219th RED HORSE Squadron," June 2, 2017.

[19] Senior Air Force leader, discussion with the authors, March 1, 2017.

[20] NCSAF, 2014, p. 28.

I-Wing pilot program would be stood up at Seymour Johnson AFB in North Carolina, and that it would be scheduled for initial operational capability in fiscal year 2017.[21] At the time this report was completed, the I-Wing pilot program had not yet been fully implemented.

### Approaches to Cross-Component Individual-Level Integration

Recent individual-level integration efforts in the Air Force include programs such as the Voluntary Extended Active Duty (VEAD) program. The VEAD program affords RC members the opportunity to serve on an extended active duty tour within an AC unit. This aids the AC in meeting force requirements, along with providing opportunities for RC members to command AC units. For instance, the current wing commander at the 325th Fighter Wing at Tyndall AFB, Florida, is an RC member who is currently on VEAD orders.[22]

Similarly, AC officers have served within RC units as wing commanders and deputy group commanders.[23] As one senior Air Force leader told us, "We have vetted folks for command billets and they consider it an honor to fill them—regardless of component."[24] The Air Force Chiefs' Group also actively considers chief master sergeants from both the AC and the RC for certain senior enlisted billets.[25] In April 2017, the Air Force also expanded its Voluntary Limited Period of Active Duty (VLPAD) program, which allows Air Force Reserve and ANG airmen from select specialties to serve on active duty in vacant active duty positions, and then return to the RC.[26]

---

[21] U.S. Department of Air Force, Secretary of Air Force Public Affairs, "AF Announces Stand Up of Integrated Wing," Washington, D.C., February 10, 2016.

[22] Phillip F. Rhodes, "Reservists Selected to Lead Active-Duty Units," *Air Reserve Personnel Center*, December 23, 2015.

[23] Secretary of the Air Force Public Affairs, 2016.

[24] Senior Air Force leader, discussion with the authors, March 1, 2017.

[25] Secretary of the Air Force Public Affairs, 2016.

[26] Kat Bailey, Air Force Personnel Center Public Affairs, "AF Adds International Affairs to VLPAD Program," San Antonio–Randolph, Tex., June 2, 2017.

*Key Takeaways*

Overall, the Air Force has a substantial history and record of developing initiatives to foster cross-component integration. One senior Air Force leader we spoke with mentioned the importance that force structure and mission sets have had in the integration of the active and reserve components in the Air Force:

> A lot of the integration piece is how we divided up mission sets across the components. For instance, 60 percent of the Air Force's aerial refueling capability is in the Guard and Reserve. When you put essential, "no fail" missions like that in the RC, it speaks to the faith you have in those components.[27]

However, as indicated in Chapter One in our discussion of the NCSAF's recommendations, perceptions remain that the Air Force could be more successful in fostering total force integration. Some individuals we spoke with also reinforced that barriers to integration remain. For instance, some Air Force leaders indicated to us that due to statutory and policy barriers, it has been easier for the AC to integrate with the Air Force Reserve than the ANG.[28] An example of such a barrier for ANG officers is that in accepting a staff tour at a major command or at the Pentagon, the officer gives up his or her spot in the state National Guard; and upon completion of the staff tour, the adjutant general of a state has to approve the officer's return to his or her respective state's National Guard. This may cause ANG officers to hesitate to take an AC staff tour.[29] As one senior leader noted, "Issues such as that make it corporately a hard chessboard to manage."[30] Many participants in our discussion also reinforced that cultural barriers to integration remain. For instance, one individual told us that "misunderstandings about the roles of the components create natural friction, and result in

---

[27] Senior Air Force leader, discussion with the authors, March 1, 2017.

[28] Senior Air Force leader, discussion with the authors, March 1, 2017.

[29] Senior Air Force leaders, discussion with the authors, November 22, 2016.

[30] Senior Air Force leader, discussion with the authors, March 1, 2017.

a bit of embedded mistrust between the AC and the RC."[31] Yet despite these remaining barriers, the Air Force continues to develop innovative approaches to cross-component integration, and the other services continue to look to the Air Force's initiatives as potential models for AC/RC integration.

### U.S. Army

Like the Air Force, the Army recently underwent a congressionally mandated force structure assessment (the NCFA). The 2016 NCFA report identified barriers to integration and offered potential recommendations the Army could implement to support total force integration. Barriers identified in the NCFA report include statutory limitations that prevent assignment of regular Army personnel into ARNG positions, inadequate billets designated for multicomponent use (both officers and NCOs), few incentives for service in MCUs, and a lack of understanding or education about the other components.[32] Our discussions with senior Army leaders highlighted these barriers, but the senior Army leaders we spoke with remained hopeful that AC/RC integration will continue to develop within the Army because there is currently strong buy-in from Army leadership to embrace a One Army/Total Force culture.[33]

Army Total Force policy is outlined in Army Directive 2012-08, *Army Total Force Policy*. Within this directive, the Army is advised to employ an integrated pay and personnel system along with "personnel policies [that] incorporate Total Force values and facilitate continuum of service and opportunities for joint experiences."[34] Total Force policy was a top priority for the previous Secretary of the Army, and it remains a top priority for the current chief of staff of the Army, General Milley. General Milley recently said that "given limited resources, we must

---

[31] Air Force leaders, roundtable discussion with the authors, February 17, 2017.

[32] See NCFA, 2016.

[33] Senior Army leaders, discussion with the authors, February 1 and 24, 2017; March 1, 2017; and April 4, 2017.

[34] U.S. Department of Army, Army Directive 2012-08, *Army Total Force Policy*, September 4, 2012.

strike the right balance of capacity and capability across the active, reserve, and National Guard forces, training and working as a team."[35] Several senior Army leaders we spoke with indicated that the current flurry of total force integration initiatives is not only in response to the NCFA recommendations, but also because current Army leadership is supportive of total force integration.[36] These initiatives include new approaches to foster cross-component integration at both the unit level and the individual level.

### Approaches to Cross-Component Unit-Level Integration

In March 2016, the Army announced that it would begin testing a new pilot program for integrating units across components called the Associated Unit Pilot Program (AUPP). This concept borrows heavily from a long-standing Air Force practice in which ANG, Air Force Reserve, and active duty airmen share responsibility for piloting and maintaining aircraft at a given base.[37] The AUPP was designed to associate Regular Army, ARNG, and USAR units so that they can train together before they deploy. The first AC and ARNG units were associated in June 2016, and the second units were formally associated in October 2016.[38] When discussing this concept before the House Armed Services Committee in 2016, General Milley said, "What we're trying to do is put teeth behind the idea of 'total force' and make that real, to walk the walk, not just talk the talk."[39]

It is important to note that the AUPP also has an element that fosters cross-component integration at the individual level. In the case of personnel exchanges between the AC and ARNG, the Secretary of the Army has to sign the memo to have AC detailees in ARNG units,

---

[35] Gen. Mark A. Milley, Chief of Staff of the Army, "Winning Matters, Especially in a Complex World," Association of the United States Army, October 5, 2015.

[36] Senior Army leaders, discussion with the authors, February 1 and 24, 2017; March 1, 2017; and April 4, 2017.

[37] Jared Serbu, "Army to Experiment with New Blended Units of Active, Reserve Forces," *Federal News Radio*, March 22, 2016.

[38] Army AUPP representative, discussion with the authors, March 2, 2017.

[39] Serbu, 2016.

and the President of the United States has to approve a list of AC officers for state commission.[40] This is another reminder of the potential statutory and policy barriers to cross-component integration, especially concerning integration across active (Title 10) and National Guard (Title 32) personnel.

### Approaches to Cross-Component Individual-Level Integration

In addition to the AUPP personnel exchanges, the Army is pursuing other changes in personnel policy to facilitate continuum of service, and it is establishing pilot programs that will incentivize cross-component assignments. For instance, the pilot Total Force Assignment Program (TFAP) will provide cross-component opportunities for junior AC captains to take a two-year assignment that includes one year of training in a USAR rapid call-up unit and another year in command of that USAR unit.[41] The design of the TFAP offers important insights for the other services because the program has deliberately tried to identify a point in AC officers' careers when they could participate in a cross-component assignment without such an assignment competing with other mandatory career requirements. The result is that the TFAP will be offered to junior AC captains who have just completed the advanced course—a time during which they would be conducting routine staff assignments. Thus, one of the major incentives for participating in the TFAP will be the opportunity for junior AC captains to take a second command assignment during a point in their career in which they otherwise would not have that opportunity.

In a roundtable discussion we had with Army leaders, this type of career development for both RC and AC officers was a primary concern when discussing cross-component assignments. This theme was primarily in reference to NCFA recommendation number 28, which recommends that "[t]he Secretary of the Army should develop selection and promotion policies that incentivize Regular Army, ARNG, and

---

[40] Title 32 U.S.C. 315 authorizes the secretaries of the Army and Air Force to detail a Title 10 officer or enlisted member to a state National Guard. However, a Title 10 officer detailed at a state National Guard may accept a state commission in ARNG or ANG only with the consent of the President and the state governor involved.

[41] Army TFAP representative, discussion with the authors, February 22, 2017.

USAR assignments across components and within MCUs."[42] One Army leader remarked, "Initially it didn't appear that the Army was in favor of recommendation number 28 because it would not provide AC members with vertical career path mobility. That's both the AC's and RC's biggest concern."[43] Another leader cautioned that integration policy should not "be so overarching that it constrains the components' ability to pick the cream of the crop for assignment."[44]

The Army has also stood up a Talent Management Task Force in the Army G-1 that is tasked with improving and maximizing human capital in the Army in order to improve readiness.[45] As part of its efforts, this task force is making great strides in addressing barriers to AC/RC integration and permeability across components. The Talent Management Task Force is also trying to identify ways to make current "one size fits all" Army human resource processes more flexible in order to facilitate individual talent management.[46] This includes developing policies to facilitate the movement across components in order to recruit and retain service members who may want flexibility in their career path. This includes the Career Intermission Program (CIP), a program in which AC members can be temporarily released from their AC assignments for an appointment in the Individual Ready Reserve for a period of up to three years while retaining their benefits. While CIP increases permeability across components, one senior Army leader told us that only seven people in the Army are participating in the program because the CIP has not been widely advertised.[47] Individual talent management and permeability should also be facilitated with the development of the Integrated Pay and Personnel System-Army (IPPS-A), which is an online human resource system that will provide integrated personnel, pay, and talent management capabilities in a single system

---

[42] NCFA, 2016, p. 65.

[43] Army leaders, roundtable discussion with the authors, April 3, 2017.

[44] Army leaders, roundtable discussion with the authors, April 3, 2017.

[45] Senior Army leader, discussion with the authors, February 24, 2017.

[46] Senior Army leader, discussion with the authors, February 24, 2017.

[47] Senior Army leader, roundtable discussion with the authors, February 24, 2017.

to members of all of the Army's components. This system will also allow commanders to identify skill sets in all of the components so that they can fill positions with personnel who have the skills they need, regardless of the component in which they reside.

### Key Takeaways

While some of the Army's integration efforts have been stymied in the past by cultural barriers across the Regular Army, ARNG, and USAR, the Army is currently making headway in the development of innovative approaches to cross-component integration at both the unit and individual levels. The AUPP, TFAP, and the Army's Talent Management Task Force are all promising initiatives that respond to the NCFA's recommendations. The AUPP is largely modeled after the Air Force's associate programs, but the TFAP and Talent Management Task Force offer important novel insights to the other services. The design of the TFAP highlights the importance of identifying when cross-component assignments can be offered to service members at a point in their career when those assignments do not compete with other career requirements and they do not harm the typical career path of participants. The Army's Talent Management Task Force highlights the need to offer more flexibility in currently rigid military human resources policies, career progression paths, and time lines. By increasing permeability across components, the military services will be able to access personnel in key capabilities (e.g., pilots, cyber experts) who may otherwise not join the military under traditional conditions of service. Since these pilot programs are new, it will be critical that the Army evaluate their success in accomplishing their goals so that they can learn from and improve them.

## U.S. Coast Guard

The Coast Guard's AC/RC integration experience is unique among the armed services, and while the Coast Guard is part of the Department of Homeland Security, not DoD, its experience provides many lessons that could inform the implementation of future DoD AC/RC integration efforts. As part of efforts to streamline the Coast Guard and decrease overhead, in 1993, the chief of staff of the Coast Guard chartered the

Reserve Field Organization Quality Action team "to recommend a service-wide standard organizational structure and support system for the Coast Guard Reserve."[48] After a study of this issue was published in April 1994, the commandant of the Coast Guard approved a course of action to integrate the Coast Guard's active and reserve components.[49] As a result of this "Team Coast Guard" effort, RC members were integrated into AC units under a single AC chain of command, and they transitioned from conducting primarily administrative duties to operational duties. The implementation of this change occurred quickly and was not without its challenges, including the loss of RC command positions and the lack of a deliberate policy for the employment and training of the Coast Guard Reserve.[50] In response to these concerns, in 2014, the Coast Guard implemented the Reserve Force Readiness System, a service-wide readiness infrastructure that matches resources with requirements, and attains and maintains readiness to facilitate the activation and deployment of the Coast Guard Reserve when surge operations require additional personnel for the AC.[51]

Today most Coast Guard reservists are "assigned to the same active duty command that they would augment upon mobilization, they are better-prepared both administratively and operationally to report, in most cases, within 24 hours of call-up."[52] Since the initial decision to integrate its AC and RC, the Coast Guard has also taken additional steps, including requiring the AC and RC to train on the same equipment so that the RC can easily augment the AC when

---

[48] John R. Brinkerhoff and Stanley A. Horowitz, *Active-Reserve Integration in the Coast Guard*, Alexandria, Va.: Institute for Defense Analyses, 1996, p. IV-9.

[49] Brinkerhoff and Horowitz, 1996, p. IV-16.

[50] Coast Guard leaders, discussion with the authors, February 1, 2017.

[51] See U.S. Coast Guard, Commandant Instruction 5320.4A, *Reserve Force Readiness System (RFRS) Staff Element Responsibilities*, November 6, 2014.

[52] U.S. Coast Guard, "Coast Guard Reserve History," December 7, 2018. Exceptions to this are the Coast Guard Port Security Units (PSUs), which are staffed almost solely with reservists, and Naval Reserve Harbor Defense Command Units, which have reservists assigned to them (U.S. Coast Guard, 2016).

necessary.[53] In this section, we provide more details of the consolidation of RC personnel into AC units, we discuss a new initiative in the Coast Guard's Office of Boat Forces to deliberately plan and manage its reserve members, and we discuss key takeaways from the Coast Guard experience. We do not discuss unit-level integration initiatives, because the Coast Guard does not integrate or associate AC and RC units—the only remaining units that are primarily composed of RC members are PSUs and Naval Reserve Harbor Defense Command Units, but they do not formally integrate with AC units.[54]

### Integration of RC Personnel into AC Units

Prior to integration of the Coast Guard's active and reserve components in 1994, the Coast Guard Reserve was responsible for administrative duties and some enlisted reservists were fully employed augmenting the mission under the authority of operational commands. This created a bit of an issue with operational commands qualifying and certifying reservists who technically belonged to a separate command. Since some reservists did not participate in day-to-day operational missions, concerns arose about their operational readiness. We were told by several senior Coast Guard leaders that during this time, morale was low in the Coast Guard Reserve because members did not feel as though they were "part of the team" and they felt disengaged from operational missions.[55] When the decision was made in 1994 to reorganize the RC and integrate its members into AC units, this represented a shift from a model in which reservists were employed as individual augmentees on an as-needed basis, to one in which RC members became part of an AC unit with which they trained and deployed. As one senior Coast Guard leader who experienced this transition told us, "It was a huge paradigm shift."[56]

One of the major unforeseen challenges associated with the integration effort of the United States Coast Guard (USCG) in the 1990s

---

[53] Senior Coast Guard officials, discussion with the authors, March 2, 2017.

[54] U.S. Coast Guard, 2018.

[55] Senior Coast Guard officials, discussion with the authors, March 2, 2017.

[56] Senior Coast Guard leaders, discussion with the authors, March 3, 2017.

was the loss of RC leadership positions because all RC personnel became integrated into AC chains of command. As a result, the Coast Guard has had to identify other opportunities in which RC members can advance their careers.[57] When we spoke with senior Coast Guard leaders, several mentioned that the limited opportunities for command positions as a reservist incentivized them to take as many active duty assignments as they could in order to advance their careers.[58] One senior leader said that he felt he was "caught in reserve world."[59] However, because many reservists have since acquired such diversified experiences, there was consensus among the individuals we spoke with that the AC now often seeks out RC members to fill positions.[60] We also heard that while integration has been effective at the field level, it never occurred at the headquarters level. This has been problematic in that the Coast Guard Reserve is not positioned to foster conversations on requirements development, enterprise risk decisions, and resourcing.

There was also consensus among the individuals that we spoke with that one of the biggest facilitators of AC/RC integration in the 1990s was the development of an IPPS.[61] Prior to those integration efforts, the Coast Guard had two separate administrative systems for pay, billets, and personnel information. In 1996, the Coast Guard instituted a single administrative system for both active and reserve component members.[62] IPPS enabled the Coast Guard to identify skill sets across both components, and from the service member's perspective, IPPS decreased the number of problems that arose with pay and benefits. However, we should note that several individuals we spoke with indicated that migration between the AC and the RC in the Coast Guard remains "clunky."[63]

---

[57] These include commanding PSUs.

[58] Senior Coast Guard leaders, discussion with the authors, March 3, 2017.

[59] Senior Coast Guard leaders, discussion with the authors, March 3, 2017.

[60] Senior Coast Guard leaders, discussion with the authors, March 3, 2017.

[61] Senior Coast Guard leaders, discussion with the authors, March 3, 2017.

[62] Brinkerhoff and Horowitz, 1996, p. IV-16.

[63] Senior Coast Guard leaders, discussion with the authors, March 3, 2017.

### Boat Forces Reserve Management Plan

While the Coast Guard Reserve became a national mobilization asset after the integration efforts of the 1990s, "an effective national management strategy was slow to follow."[64] In addition, the mobilization process, which emphasized the individual, became disconnected from the overall readiness of the force.[65] The Coast Guard's Office of Boat Forces has been at the forefront of the Coast Guard in its active management of its reserve force.[66] The Boat Forces Reserve Management Project (BFRMP) is a five-year initiative that went into effect in January 2014. It was established to set clear goals for the utilization of reservists, match reservists to unit capacity, and bring predictability to the mobilization process.[67] The BFRMP also seeks to increase readiness by requiring the same level of competency for both active and reserve component members (although given their limited training time, RC members receive more time to achieve those levels of competency). The Coast Guard has found that readiness has increased as a result—while only 13 percent of reserve forces achieved key certifications prior to 9/11, certification rates have recently increased 160 percent.[68]

The BFRMP also established a new mobilization process called the Reserve Readiness Cycle (R2C). The R2C identifies on-call boat crews that are ready to respond when a disaster hits. The crews drill together, and they know well in advance when their two-month-per-year duty period is, so their deployment preparation can be completed ahead of time.[69] In many ways, the R2C model has many similarities to the Army's Force Generation model.

---

[64] David Ruhling, "Shaping the Reserve Workforce," *Reservist*, Vol. 61, No. 1, 2014, p. 24.

[65] Ruhling, 2014, p. 24.

[66] The Coast Guard has a unique model of managing its reserve forces. Individual offices in the Coast Guard request reserve personnel from the Office of Reserve Affairs, which then assigns reserve personnel to the offices. The individual offices are responsible for managing their reserve personnel.

[67] Mark E. Butt, "The View from the Bridge," *Reservist*, Vol. 61, No. 1, 2014, p. 6.

[68] Senior Coast Guard officials, discussion with the authors, March 2, 2017.

[69] Ruhling, 2014, p. 30.

### Key Takeaways

From discussions we had with senior Coast Guard leaders, we learned that the integration effort in the late 1990s was successful in developing a stronger total force culture. Today, the USCG Reserve is integrated into active USCG day-to-day operations. Since RC personnel are now directly assigned to the active units in which they train, their efforts and skills are directly applied to the unit's mission; and by all accounts, they feel more engaged in the broader Coast Guard mission. The Coast Guard continues to develop initiatives like the BFRMP to deepen the total force culture in the Coast Guard. For instance, the Coast Guard (like the Army) is also exploring more flexible individual talent management strategies rather than managing by component. As one individual told us, "It should be talent management, not tribal management."[70]

The Coast Guard experience also offers several cautionary lessons. For instance, some observers argue that the integration took place too quickly and that there was no clear end state or structure that defined the role of the Coast Guard Reserve.[71] Ultimately this resulted in concerns over the readiness of the Coast Guard Reserve. In addition, the loss of RC command positions during the integration process also forced the Coast Guard to quickly identify other leadership opportunities for its RC members. Several individuals we spoke with cautioned not to implement far-reaching integration changes too quickly, and instead plan deliberately, identify the end state, and clarify the missions of the components.

### U.S. Marine Corps

The U.S. Marine Corps offers examples of integration at both the unit level and the individual level. At the unit level, Marine Air-Ground Task Force, a deployable structure that can vary in size and composition according to mission, can incorporate RC units and detachments within the overall structure.[72] In addition, the Marine Corps provides

---

[70] Coast Guard officials, discussion with the authors, February 1, 2017.

[71] Coast Guard officials, discussion with the authors, February 1, 2017.

[72] See Stephanie Leguizamon, U.S. Marine Corps Forces Reserve, "Reserve Marines Prove Readiness to Support the Active Component at ITX 4-17," Twentynine Palms, Calif., July 3, 2017.

robust examples of individual integration. In 2004, the Marine Corps conducted a Total Force Structure Review.[73] One result of this review was the increased use of Individual Mobilization Augmentees (IMAs) in order to facilitate timely access to reserve forces.[74] The Inspector and Instructor (I&I) program is the Marine Corps' main cross-component integration effort and is viewed by many of the other services as a model program. The I&I program is unique in that it places AC and Active Reserve personnel within Select Marine Corps Reserve (SMCR) units. Since the I&I program is the Marine Corps' signature integration program for providing regular and routine cross-component integration, we focus in this section on a more detailed discussion of the program.

### Marine Corps I&I Program

Members on I&I duty are tasked to "instruct and assist SMCR units to maintain a continuous state of readiness for mobilization; inspect and render technical advice in command functions including administration, logistical support, and public affairs; and execute such collateral functions as may be directed by higher authority."[75] One Marine Corps I&I representative told us:

> From my perspective, the I&I program ensures that the RC units are capable of deploying, because they are trained and ready to do so. A secondary effect of that is when these units do deploy, the I&I folks stay behind and run the unit while they are away, while also providing the opportunity for AC folks to have staff experience working with the RC. It gives the AC a better perspective of the RC lifestyle and requirements differences.[76]

---

[73] John W. Bergman, "Marine Forces Reserve in Transition," *Joint Force Quarterly*, No. 43, 2006, p. 27.

[74] Michael W. Hagee, Commandant of the Marine Corps, Statement to U.S. Senate Committee on Armed Services, *Defense Authorization Request for Fiscal Year 2006 and the Future Years Defense Program*, 109th Cong., 1st sess., February 10, 2005.

[75] U.S. Marine Corps Forces Reserve, "Definitions."

[76] USMC I&I representative, discussion with the authors, March 23, 2017.

In addition, this program places AC, Active Reserve, and Selected Reserve members all within a single chain of command. This is similar to the Coast Guard and some MCUs in the other services. There are, however, two different models in the I&I program—one for aviation and one for infantry.

On the aviation side, I&I personnel are referred to as "site support," and "you have the wing commander who is a RC General Officer, then below that is the Operations Group Commander who is an AC Colonel, then down to the RC squadron with a site support function."[77] At the squadron level, an RC unit is fully integrated with Selected Reserve and Active Reserve, along with a small contingent of AC personnel. The personnel makeup within an RC infantry battalion is predominantly one component, as opposed to aviation units, which have a mix of components.

On the infantry side, the battalions report to their regiments, which have AC commanders, and the regiments report to the divisions, which have an RC general officer. The division commander in turn reports to an Active Reserve general officer or an AC general officer if the commandant wishes.[78] We found that battalions aligned to Marine Forces Reserve utilize their I&I personnel in a unique way. Typically, an RC battalion will have an RC commander and an I&I Marine who serves as the de facto commander when the RC commander is out. This creates an environment in which there is continuity of command.

### Key Takeaways

The Marine Corps is often viewed both in the military and by external experts as providing examples of successful cross-component integration at both the unit level and the individual level. The I&I program in particular is seen as one of the most innovative models of AC/RC integration among the services. The model is unique because it places AC personnel in RC units—the more common form of cross-component assignment is to place RC personnel in AC units. Along these lines, the culture in the Marine Corps emphasizes that command positions

---

[77] USMC I&I representative, discussion with the authors, March 23, 2017.

[78] USMC I&I representative, discussion with the authors, March 23, 2017.

are equal, regardless of component. This culture could potentially be developed in the other services.

## U.S. Navy

In many ways, the Navy is one of the most complex services in which to implement AC/RC integration initiatives because of the Navy's missions, force structure, and equipping across its components. Unlike some of the other services, the Naval Reserve is not a mirror image of the Navy AC (e.g., there is little tactical equipment in the Naval Reserve). In addition, historically the Navy has not included its RC in strategic planning, and it has "managed its RC largely by benign neglect, because the reserve operating model simply did not fit with how the Navy operated."[79]

In the early 2000s, the Navy began to reexamine the Cold War role and structure of its RC and to identify options for AC/RC integration. The "Naval Reserve Redesign" study, completed in 2002, identified 14 specific steps to promote AC/RC integration, more than half of which had already been implemented by 2004.[80] The Navy concluded that the Naval Reserve needed to be more integrated with the AC.

### Approaches to Cross-Component Unit-Level Integration

The Navy has implemented several models of AC/RC integration at the unit level. The first of these models is the Special Capability model.[81] This model represents important niche capabilities that reside only in the Naval Reserve. The second model, Blended Units, represents a model that other services are both using and experimenting with.[82] These units have core AC personnel, but to utilize the unit's full capacity and capability requires RC members to be brought into an active

---

[79] David O. Anderson and J. A. Winnefeld, "Navy's Reserve Will Be Integrated with Active Forces," *Proceedings*, September 2004, p. 61.

[80] William A. Navas, Jr., "Integration of the Active and Reserve Navy: A Case for Transformational Change," *Naval Reserve Association News*, No. 5, 2004, p. 5.

[81] Commander, Naval Reserve Forces (COMNAVRESFOR), "Command Brief," April 22, 2016.

[82] COMNAVRESFOR, 2016.

capacity through either mobilization or a shorter-term duty status such as Active Duty for Training or Active Duty for Special Work. The final model, the Component Unit model, represents the more traditional strategic role of the RC.[83] These units mirror AC units, therefore providing the Navy with a larger mobilization capacity of like units at a lower carrying cost.

### Approaches to Cross-Component Individual-Level Integration

The Navy has a strong tradition of integration of cross-component augmentation at the individual level, which incorporates RC individuals into AC units and commands. For example, the Navy's Augmentation model consists of RC personnel that are specifically trained to provide an augmentation capability to AC Navy units.[84] These RC personnel mobilize or deploy to perform a specified mission. This available pool allows for a smaller AC force, with a ready source to draw from for manning shortfalls and additional surge capacity for operational missions or emergent crises. The Navy also has some of the most robust IMA and Full Time Support (FTS) programs among the services. IMAs are individual reservists who receive training and are preassigned to an AC billet that must be filled to meet the requirements of the AC to support mobilization (including pre- and/or postmobilization). FTS personnel perform full-time active duty service in positions that support the training and administration of the Naval Reserve Force. These types of programs have allowed the Navy to use its RC judiciously, primarily as an augmenting force when needed. This type of strategy also allows for individual AC organizations to draw on the individual skills and expertise that are resident in the RC.

### Key Takeaways

While the Navy has integrated in areas such as medical units and instillation management, the Navy's integration story offers a cautionary tale for potential future AC/RC integration efforts. Of all the services, the Navy's active and reserve components have the biggest differences

---

83  COMNAVRESFOR, 2016.

84  COMNAVRESFOR, 2016.

in their force structures and equipment.[85] For instance, tactical equipment is much more limited in the Naval Reserve (e.g., there are no aircraft carriers or submarines). These differences across the Navy's components highlight that potential DoD initiatives to further integrate the active and reserve components in the services will need to be flexible enough to account for such differences across the services. The Navy integration experience also highlights the importance of clearly defining the roles and missions of the components, and including all components in the strategic planning process.

## Common Approaches and Lessons Learned in Cross-Component Integration

In looking across the efforts of the individual military services, we found common approaches to cross-component integration at the organizational and individual levels. At the organizational level, elements of one component work with elements of a different component to perform, or train to perform, an operational mission. Examples of organizational integration include MCUs (e.g., Army and Coast Guard) and associate units (e.g., Air Force and Army).

AC and RC integration also occurs at the individual level, whereby service members from one component assume positions in another component as a unit or as part of a headquarters staff. Examples of individual integration include

- IMA programs (all services)
- AC members embedded in RC units to oversee RC readiness (e.g., Marine Corps)
- RC members embedded in a single unit under an AC chain of command (e.g., all services)
- cross-component command positions (Air Force, Marine Corps, Coast Guard).

---

[85]  Senior Navy leader, discussion with the authors, January 31, 2017.

Across the services, efforts to promote integration at the individual level typically focus on three elements of personnel management: assignments, promotion, and pay and benefits. The services all have unique challenges when it comes to incentivizing assignments to foster greater AC and RC integration. Monetary incentives for military personnel include reenlistment and retention bonuses and continuations of full-time military pay and benefits. For the RC, opportunities to remain on active duty for an extended time period are often considered a financial incentive for RC members who may be in between civilian jobs or on summer break from college, or those who aspire to add extended active duty time to their military résumés. For the AC, reenlistment and retention bonuses, such as Aviator Continuation Pay,[86] are designed to retain military personnel in career fields and occupational specialties that are often critically manned.

## Lessons Learned from Military Integration Efforts

Looking across these efforts, we identified factors that are commonly perceived as facilitating and inhibiting efforts to enhance cross-component integration. The senior leaders that we spoke with identified the following factors that could facilitate integration efforts:

- Initiatives consider unique service force structures and RC competencies.
- RC capabilities are included in service strategic planning.
- Leadership sets the tone, message, and pace regarding integration efforts.
- Initiatives define an end state and are implemented deliberately.
- Incentives are used to attract individuals to cross-component assignments (e.g., command opportunities, geographic location, and financial incentives).

On the other hand, senior leaders described the following factors as inhibiting integration:

---

[86] Aviator Continuation Pay, or Aviator Retention Pay, offers a bonus to aviation officers who agree to remain on operational flying duty for at least one year after their initial term of service. These amounts vary by service.

- cultural differences between components
- statutory and funding constraints
- lack of recognition or reward for serving in cross-component assignments
- prescriptive and rigid career development paths that inhibit cross-component talent management strategies
- lack of formal evaluation, limiting the ability to demonstrate benefits
- failures or errors in implementation.

In the next chapter, we examine the case of DoD joint integration efforts as a potential analogue to DoD's AC/RC integration efforts.

# CHAPTER THREE

# Joint Integration: A Potential Analogue for Cross-Component Integration

As part of our review of efforts to integrate various elements of the armed forces, we examined one of the major DoD reorganization efforts of the last century: the Goldwater-Nichols Department of Defense Reorganization Act of 1986. Our purpose in this chapter is to describe how this legislation was implemented and whether the mechanisms employed to foster "jointness," or cross-service cooperation, are applicable for facilitating cross-component integration.

## A Brief History of Joint Command of the U.S. Armed Forces

Following World War II, American military forces were at their zenith of performance and efficiency. The United States had just led the international military coalition that defeated both Nazi Germany and Imperial Japan in a global, joint, and combined military effort. However, despite success, America's military struggled with interoperability and jointness. World War II was characterized by global operations of individual Army and Navy commands; competition for resources and infighting were commonplace.

The idea of jointness was just beginning in World War II, with the inspiration coming from the British armed forces. The British enjoyed a highly developed staff hierarchy with the prime minister also serving as the minister of defense. His War Cabinet consisted of both the civilian and military leadership responsible for the war effort.[1] The three military members representing the British Army, the Royal Navy, and the Royal Air Force formed the Chiefs of Staff Committee. This British committee was the inspiration for U.S. President Roosevelt's new Joint Chiefs of Staff structure, meant to mirror the British command. Yet despite the progress this new organization represented, it did not guarantee smooth interoperability between the services, and America's top civilian and military leadership continued to be frustrated by what they saw as interservice rivalry and competition for resources.

Interservice rivalry continued throughout the Korean War. In 1957, four years after the Korean War, the former Supreme Allied Commander of Europe during World War II and now president, Dwight D. Eisenhower, recognized that the future of warfare was joint warfare. "Separate ground, sea and air warfare is gone forever. If ever again we should be involved in war, we will fight it in all elements, with all services, as one single concentrated effort."[2]

Despite Eisenhower's vision, the future of interoperability and jointness was a long way off. Interservice rivalry and competition for resources continued during the Vietnam War and came to an explosive head at a remote clandestine refueling site deep inside Iran, known as Desert One. After Iranian militants occupied the American embassy in Tehran and took more than 60 Americans hostage in 1979, the United States attempted a rescue mission. During refueling operations at a remote location, a helicopter and a C-130 fixed-wing aircraft collided, leaving eight dead and forcing the remaining rescuers to abort the mission.

A postmortem of the operation, known as the Holloway Report, pointed out problems with mission planning, command and control,

---

[1]   Britain's War Cabinet consisted of the foreign secretary and the minister of production, as well as the civilians who headed up the War Office, the Admiralty, and the Air Ministry. In addition, three military officers rounded out the War Cabinet: the chief of the Imperial General Staff (Army), the first sea lord and chief of the Naval Staff, and the chief of the Air Staff.

[2]   Dwight D. Eisenhower, "Address at the Centennial Celebration Banquet of the National Education Association," Washington, D.C., April 4, 1957.

and interservice operability (jointness).[3] The commission of six general and flag officers (three retired and three still serving on active duty) pointed out, "By not utilizing an existing Joint Task Force (JTF) organization, the Joint Chiefs of Staff had to start, literally from the beginning to establish a JTF, create an organization, provide a staff, develop a plan, select the units, and train the force before the first mission capability could be attained."[4] They also identified that "[c]ommand relationships below the Commander, JTF, were not clearly emphasized in some cases and were susceptible to misunderstandings under pressure."[5]

The chairman of the Joint Chiefs of Staff (CJCS) from 1978 to 1982, Air Force Gen David C. Jones, did not shy away from criticizing the system that spawned the failure at Desert One. "The corporate advice provided by the Joint Chiefs of Staff is not crisp, timely, very useful, or very influential, and that advice is often watered down and issues are papered over in the interest of achieving unanimity."[6]

Given the number of years the military had been struggling with jointness, it became clear that pressures from inside DoD were necessary, but they were insufficient on their own to push the concept of jointness to completion; only an outside catalyst would force action. As a result, to overcome inertia and skepticism and deal with armed forces with different missions, needs, and cultures, Congress was initiated to drive a new era of joint command.


## Overview of the Goldwater-Nichols Act

### Congressional Purpose and Intent

In the aftermath of failed operations resulting from a lack of jointness, Congress began to push the military toward greater cooperation and

---

3   Statement of J. L. Holloway III, "[Iran Hostage] Rescue Mission Report," August 1980.

4   Statement of Holloway, 1980.

5   Statement of Holloway, 1980.

6   William S. Lind, "JCS Reform: Can Congress Take On a Tough One?" *Air University Review*, September–October, 1985, pp. 47–50.

integration. This effort would culminate with the Goldwater-Nichols Department of Defense Reorganization Act, signed into law by President Ronald Reagan on October 4, 1986. Congressman Les Aspin, before becoming Secretary of Defense and while serving as chairman of the House Armed Services Committee, hailed the Goldwater-Nichols Act as "probably the greatest sea change in the history of the American military since the Continental Congress created the Continental Army in 1775."[7] However, the attitude was not unanimously shared. For example, ADM William J. Crowe, CJCS from 1985 to 1989, called the officer management portion of the Goldwater-Nichols Act a "horrendous case of congressional micromanagement."[8]

Congress declared eight purposes for the new law, a combination of structural changes and personnel management changes:[9]

- Reorganize DoD to strengthen civilian authority.
- Improve the military advice provided to the President, National Security Council, and the Secretary of Defense.
- Vest clear responsibility with the commanders of the unified commands and the specified command for the accomplishment of the missions assigned to their commands.
- Ensure the unified and specified combatant commanders have the full responsibility and authority to carry out their assigned missions.
- Increase attention on strategy formulation and contingency.
- Provide for a more efficient use of defense resources.
- Improve joint officer management policies.
- Enhance the effectiveness of military operations and improve DoD management and administration.

---

[7]   James R. Locher III, "Taking Stock of Goldwater-Nichols," *Joint Force Quarterly*, Autumn 1996, p. 10.

[8]   William J. Crowe, Jr., *The Line of Fire*, New York: Simon & Schuster, 1993, p. 158.

[9]   Locher, 1996, pp. 10–11.

## Structural Changes

The Goldwater-Nichols Act emphasized civilian control of the military by formalizing the chain of command from the President to the Secretary of Defense to the combatant commanders. The CJCS became the principal adviser to the President and the Secretary of Defense, and the service chiefs lost significant control and power. The purview of the service chiefs was redefined. They were now responsible for organizing, training, and equipping forces for the combatant commanders.

## Personnel Management Changes

Although the majority of changes under the Goldwater-Nichols Act (and those most visible) were structural, the personnel changes sparked by the act had a significant impact on the rank-and-file members of the military. Indeed, the Goldwater-Nichols Act did more than reorganize the organizational charts at DoD. In addition, Congress delved deep into the personnel management practices used to induce and entice officers to serve in joint assignments and to motivate the services to make their best officers available for joint assignments, thereby fostering a cultural change that would embrace jointness.

Title IV of the Goldwater-Nichols Act outlined the new approach to Joint Personnel Management.[10] It granted broad new authority to the Secretary of Defense to "establish policies, procedures, and practices for the effective management of officers of the Army, Navy, Air Force, and Marine Corps on the active-duty list who are particularly trained in, and oriented toward, joint matters."[11] Those joint matters were defined in the Goldwater-Nichols Act to mean "integrated employment of land, sea, and air forces" and explicitly included matters of national military strategy, strategic and contingency planning, and "command and control of combat operations under unified command."[12] To implement this new focus on joint matters, the act gave the Secretary of Defense the power to establish the number of officers who would serve

---

[10] Public Law 99-433, Goldwater-Nichols Department of Defense Reorganization Act of 1986, October 4, 1986.

[11] §401 of Goldwater-Nichols Act, creating §661 of Title 10, U.S.C.

[12] §401 of Goldwater-Nichols Act, creating §668 of Title 10, U.S.C.

in joint specialty positions (Joint Duty Assignment [JDA]), and created a system of education and experience requirements that were prerequisite to service as officers in joint specialties. Details were left to the Secretary of Defense, with the advice of the CJCS, but the system was to include career guidelines for "selection, military education, training, types of duty assignments; and such other matters as the Secretary considers appropriate."[13]

The creation of joint duty positions alone was no guarantee that the new concept of jointness would generate support from the military services. Officers, under the pre–Goldwater-Nichols Act system, were revered for their time and experience in service-oriented positions. Joint experience, on the other hand, was extracurricular, something that failed to resonate as desirable or in any way necessary.

Congress, through the Goldwater-Nichols Act, wanted to ensure that officers serving in joint assignments were of the same quality as those serving in the individual military services, so Title IV of the Goldwater-Nichols Act established a congressionally mandated system that reformulated officer career prospects in light of a new focus on joint personnel management. The Goldwater-Nichols Act structure was designed to position joint duty experience as a valued, desirable, and even required step for advancement to the highest echelons of the armed services. One way to do this was to compare promotion rates for three specific groups of officers—those currently "serving in" (SI) joint assignments, those who "have served" (HS) in joint assignments, and those designated as Joint Qualified Officers (JQOs)—with promotion rates of officers serving or who have served in certain in-service assignments. By law, promotion rates for these groups were periodically reported to Congress. However, there is the sense among some that "the joint force has moved beyond the point where congressional action forced it to assign quality officers to joint billets," and that "as a profession, the force has begun to manifest jointness in very principled ways."[14]

---

[13] §401 of Goldwater-Nichols Act, creating §661 of Title 10, U.S.C.

[14] Michael A. Coss, "Joint Professionals: Here Today, Here to Stay," *Joint Forces Quarterly*, No. 38, July 2005, p. 94. Also see assessments of the impacts of the Goldwater-Nichols Act

To further enforce the acceptance of joint duty as a routine part of an officer's career progression, Congress reconstituted the promotion selection boards under 10 U.S.C. 612. Under the Goldwater-Nichols Act, promotion selection boards must now contain at least one officer (chosen by the CJCS) currently serving in a joint duty assignment. Additionally, the CJCS is required to review the promotion lists to ensure that joint experience is given appropriate consideration.[15] Coupled with new guidelines (to be developed by the Secretary of Defense) for how to review and appropriately value joint experience, the post–Goldwater-Nichols Act plan for officer promotions incorporated joint duty as an equal and sometimes required set of experiences. In light of the new requirements, many officers—including those not destined for promotion to general or flag officer rank—became interested in joint duty.

## Implementation of Joint Qualification for Officers

With the significant reorganization of personnel management procedures under the Goldwater-Nichols Act, a focus on obtaining joint qualifications became very important for those seeking promotions. CJCS Instruction (CJCSI) 1330.05A, Joint Officer Management Program Procedures, contains the details on how the Goldwater-Nichols Act's new emphasis on joint qualifications was implemented. According to CJCSI 1330.05A, officers can obtain their joint qualifications one of three ways: (1) the traditional path, (2) the experiential path, or (3) a combination of the two. Traditionally, officers became JQOs after receiving joint duty credit by serving in a JDA (a specific billet or position identified on a joint manning document that provides joint experience) and completing a specific course of Joint Professional Military Education (JPME). In addition to the assignment and JPME requirement, officers were nominated for JQO status and approved by a board.

---

at the end of this chapter.

15  §402 of the Goldwater-Nichols Act.

After 9/11, personnel managers realized that an increasing number of military officers were obtaining joint experience while deployed to JTFs around the world. Military leadership believed, and Congress agreed, that these experiences should qualify officers as JQOs and a second avenue for joint qualification emerged. However, these officers still had to complete the prerequisite JPME and had to be nominated for JQO designation and approved. Later, officers were allowed to combine traditional joint assignments and experience gained while deployed to specific joint operations when competing for JQO designation. However, JPME and the nomination process were still required.

JQO designation became important to officers because it was one of the three categories of officers who received special promotion attention. Specifically, CJCSI 1330.05A required the following:

1.   Officers who are SI or HS on the Joint Staff are expected, as a group, to be promoted to the next higher grade at a rate not less than the rate for officers in the same military service in the same grade and competitive category who are currently SI or HS on the military headquarters staff (including the secretariat) of the military departments (referred to as the service headquarters average).

2.   Officers who have been designated JQO are expected, as a group, to be promoted to the next higher grade at a rate not less than the rate for officers of the same military service in the same grade and competitive category (service average).

3.   In addition to the two statutory promotion objectives above, military services will report officers who are SI or HS within the Office of the Secretary of Defense (OSD). They are expected, as a group, to be promoted at a rate not less than the rate for officers in the same military service in the same grade and competitive category who are SI or HS on their military service headquarters staff (including the secretariat) of their military service.

Promotion rates for these groups of officers were monitored by the services and reported to the Joint Staff, DoD, and eventually to Congress.

The JQO designation also became a qualification for promotion to general and flag officer because of the Goldwater-Nichols Act. This cemented the foundation of jointness. Each service now had to ensure that its best officers, those it hoped would be general or flag officers, received the right assignments so that they were JQOs before they reached consideration for their first star. This ensured the best and brightest of the services received joint experience. However, the new construct also incentivized individual officers to desire joint assignments.

## Lessons Learned from Joint Integration

There are several lessons learned that can be derived from DoD's experience with joint integration that may offer guideposts to future DoD cross-component integration efforts:

1.  Cultural change was realized. Although Congress provided the initial impetus through statutory direction, top leadership in DoD communicated the importance of and provided visible direction for these changes.
2.  Changes to assignment requirements ensured that the services sent their best individuals to joint positions.
3.  Changes to promotion requirements, requiring joint positions for promotion to senior rank, ensured that the best individuals were assigned to joint positions.
4.  Changes to training and education also ensured jointness across services. By further emphasizing joint training and education across the services, individuals became increasingly familiar with the capabilities, cultures, and processes in the other services, fostering a better understanding of a broader DoD culture.
5.  Incentives at both the service level and the individual level were automatically incorporated into the changes to the assignment, promotion, and training and education requirements that were ushered in by Goldwater-Nichols.

6.   Assignment, promotion, and training and education requirements worked together with incentives as a system to promote organizational objectives and to create an environment in which serving in joint assignments was viewed as a necessary and desirable part of an officer's career development.

Ten years after the signing of the Goldwater-Nichols Act, former Secretary of Defense Richard "Dick" Cheney said that "significant progress has been made. . . . I think Goldwater-Nichols gave it a major push."[16] When discussing the joint staff, Cheney commented on the quality of the post–Goldwater-Nichols Act staff, "[T]he policy we established requiring service on the Joint Staff [the requirement is actually to serve in joint positions] prior to moving into senior leadership positions turned out to be beneficial. We did not want anyone on the Joint Staff who did not have significant prospects back home in their own service."[17] Cheney also credited the success in Desert Storm to changes in the military structure under the Goldwater-Nichols Act. Clearly the former Secretary of Defense and future Vice President of the United States believed the Goldwater-Nichols Act had a positive impact on the services.

Yet, as with the initial efforts to enact the Goldwater-Nichols Act, its lasting legacy has been equally divisive. In November 2009, author Brad Amburn interviewed a number of senior military officers concerning the Goldwater-Nichols Act. Air Force Gen Charles "Chuck" Boyd, former deputy commander of U.S. European Command, feared that the Goldwater-Nichols Act's impact was to "politicize and de-professionalize the officer corps."[18] He believed these impacts were "unintended, but . . . real."[19] General Boyd criticized the Goldwater-Nichols Act, saying, "One effect of the legislation is that young

---

[16] "About Fighting and Winning Wars," an interview with Dick Cheney, *Proceedings*, May 1996, p. 32.

[17] "About Fighting and Winning Wars," 1996, p. 32.

[18] Brad Amburn, "The Unbearable Jointness of Being," *Foreign Policy.com*, November 16, 2009.

[19] Amburn, 2009.

officers came to believe that ticket punching was more important than anything else—that where they served was more important than what they did."[20] General Boyd believed that interservice rivalry under the Goldwater-Nichols Act was not worse, but certainly was not better. Despite the Goldwater-Nichols push to foster interservice cooperation, General Boyd felt that competition between the services was sound and healthy.[21]

Yet, in the same interview, ADM William Owens, former vice chairman of the JCS, tended to minimize the problems with Goldwater-Nichols, saying, "[T]he good it has accomplished is much more important."[22] In addition, MG William Nash, former commander of the storied 1st Armored Division, pointed out that "provisions of Goldwater-Nichols require certain assignments of service personnel in order to achieve flag rank."[23] According to MG Nash, this drives rapid turnover.[24]

Nonetheless, Secretary of Defense Ashton Carter believed "the pendulum between service equities and jointness may have swung too far. . . ."[25] Secretary Carter highlighted a major aspect of the Goldwater-Nichols Act: requiring joint duty for all officers who are promoted to general or flag rank. He said this requirement

> led to great advances in jointness across the military services—such that almost all our people know why, and how, we operate as a joint team—and it's also significantly strengthened the ability of our Chairmen, our Joint Chiefs, and our Combatant Commanders to accomplish their mission.[26]

---

[20] Amburn, 2009.

[21] Amburn, 2009.

[22] Amburn, 2009.

[23] Amburn, 2009.

[24] Amburn, 2009.

[25] Ashton Carter, "Remarks on 'Goldwater-Nichols at 30: An Agenda for Updating,'" speech, Center for Strategic and International Studies, Washington, D.C., April 5, 2016.

[26] Carter, 2016.

Secretary Carter proposed a number of changes related to joint duty. In an April 5, 2016, speech to the Center for Strategic and International Studies, he proposed reducing the time required to accumulate joint duty, from three years to two years. This offered more flexible options "to take on command assignments and other opportunities to broaden and deepen their careers."[27] Most importantly, Carter pointed out that "Goldwater-Nichols took four years to write, and it's been incredibly successful over three decades—to the credit of the reforms it put in place, we are not driven today by a signal failure like Desert One."[28]

---

[27] Carter, 2016.

[28] Carter, 2016.

# CHAPTER FOUR

# Insights from Other Integration Efforts

In seeking lessons learned from entities beyond the U.S. military that have undertaken efforts to enhance integration, RAND analyzed relevant literature, the integration and rotational programs implemented by civilian U.S. government agencies and the private sector, and the integration experiences of foreign militaries. Although the cultures and missions of these other organizations all vary and are not exact analogues to the U.S. military, their experiences offer useful insights into ways to achieve greater integration across large, and sometimes disparate, organizations. This chapter begins by drawing insights from the research and business literatures on organizational integration. Next, it provides an overview of integration efforts by U.S. nonmilitary government agencies, the private sector, and foreign militaries. Last, the chapter identifies approaches and the incentives used in these integration efforts, and the lessons learned from these integration efforts.

## Insights from the Research and Business Literatures on Organizational Integration

We began our analysis of non-DoD integration efforts by conducting a broad analysis of the literature on cultural and organizational integration, and mergers and acquisitions. We then focused on two areas that provided the most fruitful insights for DoD efforts to further facilitate integration of the active and reserve components: (1) barriers to organizational integration and ways to overcome them and (2) factors associated with successful mergers and acquisitions. Below, we discuss our findings from these sets of literatures.

### Barriers to Organizational Integration and Strategies to Overcome Them

Our review of the literature on cultural and organizational integration identified common barriers to the integration of organizations with multiple components:

- cultural mistrust/misunderstanding[1]
- resistance to change[2]
- territoriality[3]
  - concern over prestige, resources, component equities
- lack of communication and interaction across groups[4]
- unclear goals and messaging[5]
- overspecialization leading to different goals across groups.[6]

The literature also identifies several common strategies that can be used to overcome the barriers listed above:

---

[1] Colleen Lucas and Theresa Kline, "Understanding the Influence of Organizational Culture and Group Dynamics on Organizational Change and Learning," *The Learning Organization*, Vol. 15, No. 3, 2008, pp. 277–288; Rosabeth Moss Kanter, Barry A. Stein, and Todd D. Jick, *Challenge of Organizational Change: How Companies Experience It and Leaders Guide It*, New York: Free Press, 1992.

[2] Shaul Oreg, "Personality, Context and Resistance to Organizational Change," *European Journal of Work and Organizational Psychology*, Vol. 15, No. 1, 2006, pp. 73–101; Lisa Quast, "Overcome the 5 Main Reasons People Resist Change," *Forbes*, November 26, 2012; Peter M. Senge, *The Fifth Discipline: The Art and Practice of the Learning Organization*, New York: Doubleday, 1990.

[3] See Graham Brown, Thomas B. Lawrence, and Sandra L. Robinson, "Territoriality in Organizations," *Academy of Management Review*, Vol. 30, No. 3, 2005, pp. 577–594.

[4] Sugandh Kansal and Arti Chandani, "Effective Management of Change During Merger and Acquisition," *Procedia Economics and Finance*, Vol. 11, 2014, pp. 208–217.

[5] See Katinka Biljsma-Frankem, "On Managing Cultural Integration and Cultural Change in Mergers and Acquisitions," *Journal of European Industrial Training*, Vol. 25, Nos. 2/3/4, 2001, pp. 192–207.

[6] See John O'Shaughnessy, *Patterns of Business Organization*, London: Routledge, 2013.

- Top-down mandated integration can be effective in overcoming resistance to change, territoriality, and cultural distrust.[7]
- Strategies to overcome barriers include gradualism, education, and communication; participation and involvement; negotiation and agreement; burden sharing; manipulation and co-option; explicit and implicit coercion; divide and conquer; and buy-out.[8]
- Increased communication and cross-cultural interactions across groups are seen as important to developing trust.[9]
- Establishment of common culture, goals, and values fosters integration.[10]
- Integration can be facilitated through incentives (e.g., financial, career-enhancing, prestige).[11]
- Some standardization of processes can facilitate integration.[12]

---

[7] See John P. Kotter, *A Force for Change: How Leadership Differs from Management*, New York: Free Press, 1990; John W. Moran and Baird K. Brightman, "Leading Organizational Change," *Journal of Workplace Learning*, Vol. 12, No. 2, 2000, pp. 66–74; see W. Henry Lambright, "Leadership and Change at NASA: Sean O'Keefe as Administrator," *Public Administration Review*, March/April 2008, pp. 230–240; see Sergio Fernandez and Hal G. Rainey, "Managing Successful Organizational Change in the Public Sector," *Public Administration Review*, March/April 2006, pp. 168–176.

[8] See Daniel T. Holt, Achilles A. Armenakis, Hubert S. Field, and Stanley G. Harris, "Readiness for Organizational Change: The Systemic Development of a Scale," *Journal of Applied Behavioral Science*, Vol. 43, 2007, pp. 232–255.

[9] See N. DiFonzo and P. Bordia, "A Tale of Two Corporations: Managing Uncertainty During Organizational Change," *Human Resource Management*, Vol. 37, 1998, pp. 295–303. L. K. Lewis and D. R. Seibold, "Reconceptualizing Organizational Change Implementation as a Communication Problem: A Review of Literature and Research Agenda," in M. E. Roloff, ed., *Communication Yearbook 21*, Beverly Hills, Calif.: Sage, 1998, pp. 93–151; see D. M. Schweiger and A. S. Denisi, "Communication with Employees Following a Merger: A Longitudinal Field Experiment," *Academy of Management Journal*, Vol. 34, No. 1, 1991, pp. 110–135.

[10] See Susan Cartwright and Cary L. Cooper, "The Role of Culture Compatibility in Successful Organizational Marriage," *Academy of Management Perspectives*, Vol. 7, No. 2, 1993, pp. 57–70.

[11] See P. B. Clark and J. Q. Wilson, "Incentive System: A Theory of Organization," *Administrative Science Quarterly*, Vol. 6, 1961, pp. 129–166; J. Q. Wilson, *Bureaucracy: What Government Agencies Do and Why They Do It*, New York: Basic Books, 1989.

[12] See Henri Barki and Alain Pinsonneault, "A Model of Organizational Integration, Implementation Effort, and Performance," *Organization Science*, Vol. 16, No. 2, 2005, pp. 165–179.

Seen as seminal research in the field, Beckhard and Harris argued that all three of the following components must be present to overcome the resistance to change in an organization: (1) dissatisfaction with the present situation, (2) a vision of what is possible in the future, and (3) achievable first steps toward reaching this vision.[13]

### Factors Associated with Successful Mergers and Acquisitions

We also examined the literature on mergers and acquisitions to identify personnel practices that help make mergers and acquisitions successful. Marc Epstein identifies five drivers of successful postmerger integration: (1) a coherent integration strategy that "reinforces that this is a 'merger of equals' rather than an acquisition"; (2) a strong integration team that has representatives from all of the integrating organizations and that is focused on the integration, especially on eliminating any culture clashes in the integrated organization; (3) communication from senior management that is "significant, constant, and consistent," that builds confidence in the integration purpose and process, that reinforces the purpose of the integration "with a tangible set of goals," and that addresses important issues such as personnel retention and separation policies; (4) speed in implementing the integration, which will reduce uncertainty and instability; and (5) measures of success that are aligned with the strategy and vision of the integration.[14] Epstein also argues that "in personnel decisions, employees of both companies must be judged by the same standards and the candidate selection process based on merit rather than as a basis for a power struggle."[15]

The literature also identifies personnel policies that can facilitate merger and acquisition success. For instance, one of the lessons that de Noble, Gustafson, and Hergert identify for postmerger success is to cross-fertilize management teams. They point out that "whenever a merger occurs, there is a psychological hurdle to surmount in estab-

---

[13] R. Beckhard and R Harris, *Organizational Transitions: Managing Complex Change*, 2nd ed., Reading, Mass.: Addison-Wesley, 1987.

[14] See Marc Epstein, "The Drivers of Success in Post-Merger Integration," *Organizational Dynamics*, Vol. 33, No. 2, May 2004, pp. 176–179.

[15] Epstein, 2004, p. 176.

lishing a new corporate identity. It is critical to replace the 'us' vs 'them' mentality with a spirit of teamwork."[16] Peter Drucker has also suggested that during the first year of a merger, it is essential that a large number of people in the management groups of both companies receive substantial promotions.[17] Additional research indicates that human resource issues occur at several phases of the merger and acquisition process. These human resource issues include (1) retention of key talent, (2) communications, (3) retention of key managers, and (4) integration of corporate cultures.[18] We next turn to an overview of integration experiences in U.S. civilian government agencies, the private sector, and foreign militaries.

## Integration Experiences in U.S. Civilian Government Agencies

In seeking lessons learned from entities that have undertaken efforts to enhance integration, RAND analyzed the integration and rotational programs run by civilian U.S. government agencies. While we initially conducted a broad analysis of U.S. civilian government agencies, Appendix A presents a more detailed analysis of three organizations (Office of the Under Secretary of Defense [Policy], the Defense Intelligence Agency [DIA], and the Department of State [DoS]) that were selected due to their diversity of experience and workforce requirements, variation in program design, experience in using these programs to source hard-to-fill positions, and overlap with promotion processes.

Several factors motivate civilian U.S. government agencies toward greater integration. Some of these factors include the need to align

---

[16] Alex F. de Noble, Loren T. Gustafson, and Michael Hergert, "Planning for Post-Merger Integration—Eight Lessons for Merger Success," *Long Range Planning*, Vol. 21, No. 4, August 1988, p. 83.

[17] Drucker, 1981 as cited in de Noble, Gustafson, and Hergert, 1988, p. 83.

[18] See Randall Schuler and Susan Jackson, "HR Issues and Activities in Mergers and Acquisitions," *European Management Journal*, Vol. 19, No. 3, 2001, pp. 239–253; A. Charman, *Global Mergers and Acquisitions: The Human Resource Challenge*, Alexandria, Va.: Society for Human Resource Management, 1999.

efforts across divisions, to maximize resources within budgetary constraints, to accommodate cultural shifts toward greater collaboration, and to address requirements to fill critical positions. Some U.S. civilian government agencies' rotational programs emphasize intra-agency organization, while others participate in rotational programs that emphasize experience outside of one's home agency. For example, within the Intelligence Community (IC) there is a program run by the Office of the Director of National Intelligence (ODNI) called the Joint Duty Program (JDP). The JDP was created to increase integration among intelligence agencies and to increase awareness of how other agencies approach problems. After the attacks of September 11, 2001, the IC focused on incentivizing information-sharing and collaboration among the agencies, which had not been customary prior to 9/11.[19] Participants in the JDP are detailed on JDAs that last a year or more. IC personnel apply to positions, and the agencies involved in the JDA draft a memorandum of agreement governing the detail. IC personnel are incentivized to participate in the JDP since some intelligence agencies require joint duty credit in order to promote to certain senior levels. Per ODNI policy, Senior Executive Service (SES) selection across the IC requires joint duty experience.[20]

Another government agency with law enforcement responsibilities designed its internal rotational program on the premise that all its lines of effort support one integrated mission, so all of the agency's employees must have an understanding of each division within the agency. This organization's special agents must spend several years working in the agency's investigations division before assignment to their primary duty post. As a special agent gains seniority in rank, he or she would traditionally rotate throughout a series of field, supervisory, and analytic positions before being considered for an SES position. A representative of this agency explained that the reasoning for requiring these broadening rotations is so the individual's expertise is maximized across all aspects of the agency's responsibilities and he or

---

[19] JDP manager at an intelligence agency, discussion with the authors, September 23, 2016.

[20] ODNI, "Joint Duty."

she can better supervise others and coordinate effectively with other offices within the agency.[21]

The structures of rotational assignments differ significantly throughout the U.S. government. In some cases, rotations are highly encouraged, but in others, rotations are mandatory in order to promote or stay within the organization. The majority of the integration rotation-based programs sponsored by individual agencies are at the nascent stage, and many are aimed at the GS-14 and GS-15 (or equivalent) levels. In some programs, positions are held for employees while they pursue an external assignment, whereas in others, particularly when the workforce is expected to change positions every few years, employees must apply to a new position upon return to their home organization.

## Integration Efforts in the Private Sector

We found that the use of rotational programs is very common in the private sector. In fact, in many companies (particularly high-tech and engineering companies), new employees are required to spend time rotating throughout the company before they choose an area of the organization in which to work. For instance, Boeing offers early-career professionals two-, three-, and four-year rotational programs to build strong skills.[22] Intel also offers rotational programs. Its U.S. Finance Rotation allows participants to change positions every 18–36 months.[23] Some companies also require rotations in order for employees to move into upper-level positions. One of the Air Force leaders we spoke with reinforced this point when he mentioned, "I used to work for Lockheed Martin and in order to progress/climb the corporate structure you had to rotate among business areas."[24]

The private sector has found that such rotational assignments allow individuals to get a sense of the whole enterprise, allow for the

---

[21] Law enforcement agency representative, discussion with the authors, August 1, 2016.

[22] Boeing, "Rotational Programs," *Boeing*.

[23] Intel, "Rotation Program," *Intel*.

[24] Air Force leaders, discussion with the authors, February 17, 2017.

flexible employment of staff, and help prevent job burnout. Some studies have found that such rotational programs can increase retention.[25] Studies also indicate that early-career managers are more interested in rotational assignments than later-career managers, and higher-performing managers take on more rotational assignments.[26]

In addition, some private sector companies and many universities offer employees midcareer gap years or sabbaticals to broaden their careers. These types of strategies are also used as career-broadening opportunities, as well as ways to prevent job burnout among employees.[27] They offer employees opportunities to take a reprieve from their usual daily tasks to focus on the development of new skills and the acquisition of new knowledge without falling behind in their promotion gates.[28] These types of personnel strategies also benefit the organization as a whole because the employee returns after the gap year or sabbatical with new skill sets.

## Integration Experiences in Foreign Militaries

Our analysis of the integration experiences of foreign militaries focused on major U.S. allies, specifically Australia, Canada, and the United Kingdom. It is important to note that these militaries are smaller than the U.S. military and that they have different missions and force structures. However, these militaries may provide insights into integration efforts because they too are in the process of identifying ways to further integrate their active and reserve components.

---

[25] National Association of Colleges and Employers, "Rotational Programs Yield Higher Retention Rates," March 22, 2017.

[26] See Lisa Campion, "Study Clarifies Job-Rotation Benefits," *Workforce*, November 1, 1996; Steven G. Rogelberg, *The SAGE Encyclopedia of Industrial and Organizational Psychology*, Thousand Oaks, Calif.: Sage Publications, 2007.

[27] David Burke, "The Surprising Benefit of Work Sabbaticals," *Forbes*, June 29, 2016.

[28] Elizabeth Garone, "The Surprising Benefits of a Mid-Career Break," *BBC*, March 28, 2016.

It is also important to note that the size and status of RCs in many foreign militaries have changed considerably over the last 20–30 years. Most notably, reserve forces across Europe have significantly declined in size.[29] For instance, in 2012, NATO members such as Belgium, Italy, and the Netherlands had less than 10 percent of their Cold War reserves.[30] Non-NATO countries such as Sweden and Switzerland had one-third of the reserve forces that were available in 1990.[31]

In addition, the role of RCs in foreign militaries has also evolved over time:

> Nations no longer consider their reservists as strategic assets suitable primarily for mobilization during major wars. Whereas previously they managed reservists as supplementary forces for use mainly during national emergencies, major governments now increasingly treat reservists as complementary and integral components of their "total" military forces.[32]

Consequently, the major military powers have widely adopted "total force" policies that treat their active and reserve components as integrated if not totally interchangeable elements. Government policies increasingly treat mobilized reservists and regular forces similarly—harmonizing their organizational structures, compensation packages, and rules and regulations—as they link the two components more tightly.[33]

---

[29] Timothy Edmunds, Antonia Dawes, Paul Higate, K. Neil Jenkings, and Rachel Woodward, "Reserve Forces and the Transformation of British Military Organisation: Soldiers, Citizens and Society," *Defence Studies*, Vol. 16, No. 2, 2016, p. 120.

[30] "Send in the Reserves," *Armed Forces Journal*, February 1, 2012.

[31] "Send in the Reserves," 2012.

[32] Richard Weitz, *The Reserve Policies of Nations: A Comparative Analysis*, Carlisle, Pa.: Strategic Studies Institute, September 2007, p. vii.

[33] Weitz, 2007, p. viii.

## Australia

Until recently, the Australians saw their RCs as primarily a homeland defense force.[34] However, given the large role that the Australian Defence Force (ADF) has played in the conflicts in Iraq and Afghanistan, the perceived role of the RCs has changed. Research and feedback from over 10,000 ADF personnel revealed that active personnel members want more flexibility and reservists are seeking more opportunities to serve.[35]

As a result, ADF is undergoing a major effort to integrate its active and reserve components, and to facilitate permeability across components so that it can better harness skills and expertise across the components. This effort is called Project Suakin. The main outcome of the plan is to "develop a contemporary employment model with associated conditions of service based on the concept of an ADF career for life, particularly to allow permanents to move seamlessly to part time work in their work life balance and for reservists to move seamlessly to full time work in the ADF."[36] In conjunction with Project Suakin, the Australian military is also implementing a "Total Workforce Model" that offers improved access to flexible career options by enabling mobility across full-time and part-time service as personal circumstances change.[37] As these novel total force personnel management approaches continue to be implemented by the Australian military, the U.S. military should observe how the implementation process unfolds, whether the U.S. military can learn from these new policies and practices, and whether they might be applicable to U.S. AC/RC integration efforts.

## Canada

In accordance with a recommendation of a 1987 White Paper, Canada adopted a total force principle to govern the integration of its active

---

[34] Weitz, 2007, p. 71.

[35] Australian Government, Department of Defence, "ADF Total Workforce Model."

[36] Defence Reserves Association Submission—Defence White Paper 2015, p. 11.

[37] Australian Government, Department of Defence, "ADF Total Workforce Model."

and reserve components. The most comprehensive analysis of Canada's RCs is the 2000 Fraser Report, formally titled *In Service of the Nation: Canada's Citizen Soldiers for the 21st Century*.[38] In the report, John A. Fraser, the chairman of a special committee charged with assessing the state of the country's RCs and policies, highlighted continued problems in the training of Army reservists. In particular, the committee found that "although the Army leadership had tried to create 'reserve friendly' training packages, part time soldiers could rarely achieve the same standards as full time professionals."[39]

Since the publication of the Fraser Report, the Canadian Forces have undertaken several projects to address personnel policy shortfalls that inhibit cross-component service. For instance, the Air Force has developed a formal policy of facilitating transfers between its Reserve and Regular components. In recent years, it has also adopted measures to harmonize career policies that previously restricted movement between them.[40] In addition, Canada has also made training requirements the same across its ACs and RCs to facilitate utilization of personnel across components.

## United Kingdom

To date, the British military's attempts at a Total or "Whole Force" have not been fully successful, and progress toward a whole force has varied between the services.[41] As in many other countries, there are two different views about the reserve forces in the United Kingdom (UK). One side takes the view that "the reserves ought to be smaller and integrated with the active force, while the other side sees the reserves as a somewhat larger force with a different role, taking on those tasks that a small active force cannot do."[42] The UK Ministry of Defence has tried to arrange for reservists to spend at least some time training with

---

[38] Weitz, 2007, p. 66.

[39] Weitz, 2007, p. 66.

[40] Weitz, 2007, p. 60.

[41] Mark Phillips, *The Future of the UK's Reserve Forces*, London: RUSI, April 2012, p. vi.

[42] "Send in the Reserves," 2012.

the regular units they would join on deployments, but such integration has not always proved possible.[43]

The UK's "Future Reserves 2020: Delivering the Nation's Security Together" initiative moves toward fully integrating the UK military's active and reserve components.[44] The Future Reserves 2020 initiative seeks to move more specialties into the RC, and it sets out to recruit 30,000 reservists by 2020—in part to make up for cuts to regular forces.[45] However, the UK is having difficulty reaching that recruiting goal. While the Future Reserves 2020 report states that "closer integration of Reservists within a wider range of challenging tasks develops a culture of mutual respect between the Regular and Reserve components,"[46] and that "the Whole Force concept must be at the very heart of how the Ministry of Defense manages its human capability," it does not suggest specific personnel strategies in which such integration could be facilitated.[47] However, it does call for the modernization of equipment so that AC and RC forces train on the same equipment.

## Approaches Found in Other Organizations' Integration Efforts

In looking across our analysis of the integration efforts of U.S. civilian agencies, the private sector, and foreign militaries, we identified approaches to integration that could potentially be applied in the U.S. military to further integrate the active and reserve components. These approaches are listed in Figure 4.1.

---

[43] Weitz, 2007, p. 31.

[44] UK Ministry of Defence, *The Independent Commission to Review the United Kingdom's Reserve Forces*, London, July 2011; UK Ministry of Defence, "Consultation Launched on the Future of Britain's Reserve Forces," November 8, 2012.

[45] Edmunds et al., 2016, p. 119.

[46] UK Ministry of Defence, 2011, p. 21.

[47] UK Ministry of Defence, 2011, p. 44.

**Figure 4.1**
**Approaches Found in Other Organizations**

| Civilian Agencies | Private Sector | Foreign Militaries |
|---|---|---|
| • Link cross-component assignments to future assignments<br>• Give points for promotion for cross-component assignments<br>• Provide an additional retirement annuity benefit for cross-component assignments<br>• Offer early bidding for next assignments after cross-component assignments<br>• Make additional education and training opportunities available after cross-component assignments<br>• Provide choice in duty location after cross-component assignments<br>• Mandate cross-component assignments for promotion | • Use rotational assignments to expand individuals' knowledge of broader enterprise operations<br>• Require new hires to rotate around the organization before deciding which component to work in<br>• Offer midcareer gap years or sabbaticals to broaden their careers | • Increase permeability across components to harness skills and expertise<br>• Change conditions of service to allow for more flexibility across components<br>• Make training requirements the same across AC and RC to facilitate utilization of personnel across components<br>• Train on the same equipment, regardless of component |

These approaches aim to foster integration by offering opportunities for career-broadening experiences, standardize training across organizations, and offer more flexibility to employees in their career progression paths. These types of approaches align with the direction in which some of the services are already headed. While current personnel policies in some of the services are too rigid to accommodate some of the approaches above, the current trends toward permeability and individual talent management open the door for more flexible personnel management approaches to AC/RC integration.

Incentives to pursue rotational assignments vary depending on the organization and the specific nature of the assignment. However, our findings from our discussions indicate that within civilian government programs, financial incentives tend to be less important to employees than other incentives that offer work-life balance or greater career enhancement potential in the form of promotions or future assignment preferences. People we spoke with also noted that incentive programs that suffer from lack of clarity or in uneven application will depress participation in the program.

## Crosscutting Lessons Learned from These Integration Efforts

When looking across the experiences of U.S. civilian agencies, the private sector, and foreign militaries, we can draw several crosscutting lessons:

- Fostering integration requires a shift in culture and leadership buy-in.
- Required rotations can improve retention and facilitate a holistic understanding of enterprise operations.
- Integration is often easier with more junior employees.
- Financial incentives are not always the most compelling.

We discuss each of these lessons learned below.

### Fostering Integration Requires a Shift in Culture and Leadership Buy-In

Offices within civilian U.S. government agencies often exhibit pronounced cultural attributes that can be challenging to integration, such as insularity stemming from protection of information. These attributes can be very difficult to influence, particularly if promotion or other career advancement is predicated on adherence to these norms. Without leadership commitment and clear direction to change incentives, promotion structures, or integration programs, cultural norms are likely to subsume integration efforts. DIA in particular highlighted the importance of leadership buy-in and endorsement, which was critical in DIA's case to implementing internal rotation programs.

## Required Rotations Can Improve Retention and Facilitate a Holistic Understanding of Enterprise Operations

Our analysis of the policies and practices of other government agencies and the private sector indicates that required rotation/cross-component assignment can help improve retention and facilitate a holistic understanding of an organization's operations. The literature and our discussions with representatives from other government agencies indicate that these types of assignments help improve retention because they help broaden employees' skill sets and help prevent employee burnout. Many companies in the private sector also require such rotations at the beginning of an employee's career so that the employee can develop an understanding of what different parts of the organization do. Employees do not decide which part of the organization they want to work in until after several rotations have been completed. Some companies have found that this type of strategy ultimately increases retention because employees decide which part of the organization is the best fit for them, and it helps grow employees who have a broader understanding of the organization's different components and their different competencies.

Frequently, senior-level and management positions require awareness of a broad range of agency (and interagency) responsibilities that can be gained via hands-on experience within different offices. DoS's Foreign Service program in particular approaches career development by emphasizing various tours and rotations throughout DoS, beginning at the entry level. Making these rotations a requirement creates clear expectations and benchmarks for all its employees and facilitates transparency in its promotion process.

## Integration Is Often Easier with More Junior Employees

The millennial generation tends to favor transparent coordination, information-sharing, collaboration, and rapid feedback.[48] Starting integration programs from the ground up is a practice also favored by many

---

[48] Carol A. Martin, "From High Maintenance to High Productivity: What Managers Need to Know About Generation Y," *Industrial and Commercial Training*, Vol. 37, No. 1, 2005, pp. 39–44.

organizations we spoke with (including the United States Air Force and the United States Coast Guard). It not only indoctrinates principles of open collaboration from the onset of one's career but also focuses on a younger generation that is predisposed to being eager to work with others to achieve solutions.[49] DIA in particular emphasized this point, noting that many of its younger employees were drawn to it for its ability to offer employees the chance to mobilize in support of U.S. operations overseas and work alongside other agencies and parts of the U.S. DoD.

### Financial Incentives Are Not Always the Most Compelling

Our analysis found that incentives that are tied to promotion, follow-on assignments, or work-life balance are often more compelling than additional types of financial incentives beyond those that accompany promotions. Examples of these additional types of financial incentives include bonuses and tax-free benefits. For instance, rotational programs in the Office of the Under Secretary of Defense (Policy) (OUSDP), DIA, and DoS are all tied to career-enhancing motivating factors. While certain financial benefits are attached to certain rotational assignments (such as tax-free allowances), these programs emphasize other incentives above financial ones. Further, both DIA and DoS require broadening rotational assignments in order to secure promotion at more senior GS levels. The employees that each of these agencies attract may tend to be service-focused and motivated by mission rather than financial incentives, which could provide a useful analogue to military motivations.

---

[49] Deloitte, *Big Demands and High Expectations: The Deloitte Millennial Survey—Executive Summary*, New York, January 2014, p. 3.

# Applying Findings to Better Facilitate AC/RC Integration

In thinking about how to apply the findings in Chapters Two, Three, and Four to identify ways in which the services can better facilitate AC/ RC integration, we developed a multifaceted strategic human resources framework based on commonly used strategies from our findings that can facilitate integration and achieve organizational goals. As indicated in Figure 5.1, this integrative framework is composed of five

**Figure 5.1**
**Multifaceted Framework for Increasing AC/RC Integration**

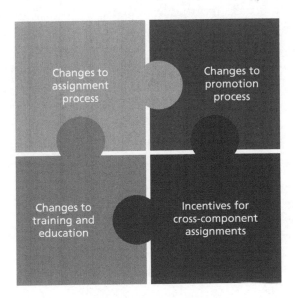

facets: (1) changes to assignment processes, (2) changes to promotion processes, (3) changes to training and education, (4) incentives for cross-component assignments, and (5) structural changes (e.g., legal, regulatory, systems changes).

Within each of the elements in this framework, we identified potential actions that DoD could take to increase AC/RC integration. These potential actions are summarized in Figure 5.2.

In the remainder of this chapter, we describe in detail the potential actions that could be taken with regard to each element of our framework to facilitate further AC/RC integration, and we discuss considerations regarding when to offer service members cross-component programs.

## Changes to Assignment Processes

### Identify and Expand Positions Suitable for Cross-Component Assignments

One of the first steps that the services could take is to identify and expand the number of positions that are suitable for cross-component assignments, and create models of integrated cross-component force structure. As indicated in Chapter Two, the Goldwater-Nichols Act established the number of officers who would serve in joint duty assignments, and it created a system of education and experience requirements that were prerequisite to service as officers in joint specialties. A similar set of requirements could potentially be established for cross-component assignments. However, it is important to note that during our discussions with senior leaders and other representatives of the services, many expressed hesitations to create additional "cross-component" requirements since it is already so difficult for service members to meet all of their career requirements in order to advance.

### Improve Screening for Cross-Component Assignments to Ensure High-Quality Candidates

Our findings from Chapters Three and Four indicate that some previous integration efforts failed because the services did not adequately

**Figure 5.2**
**Summary of Potential Actions to Increase AC/RC Integration**

**Changes to Assignment Processes**

- Identify and expand positions suitable for cross-component assignments
- Improve screening for cross-component assignments to ensure high-quality candidates
- Utilize potential incentives as needed to fill cross-component assignments
- Consider and mitigate effects of changes on AC and RC career paths

**Changes to Training and Education**

- Expand opportunities for cross-component training

**Changes to Promotion Processes**

- Clarify precepts and board changes
- Expand board membership
- Develop AC/RC qualification system

**Incentives for Cross-Component Assignments**

- Identify appropriate combination of monetary and nonmonetary incentives
  Monetary incentives include:
  o assignment and incentive pays
  o subsidies for housing
  o subsidies for childcare and family benefits
  Nonmonetary incentives include:
  o cross-component assignments linked to future assignments
  o work-life balance
  o award ribbon or qualification for cross-component assignments

**Structural Changes**

- Ensure Total Force strategy guides integration efforts
- Implement talent management workforce strategy
- Address legal/regulatory challenges and undertake efforts to develop mitigation strategies
  o Duty status reform
  o Improve scrolling process (the transfer of a service member from one component to another)
- System changes (e.g., fully implement Integrated Pay and Personnel Systems)

screen for high-quality candidates to participate in those efforts. There was agreement among the people we spoke with across the services that in order to maximize the success of any new cross-component integration programs, high-quality candidates must be chosen to participate in them. Otherwise, the programs may fail due to no other reason than the participation of low-performing candidates. Candidates could be screened in multiple ways, including setting rigorous minimum criteria for cross-component assignments and providing incentives for high-quality candidates to apply to cross-component assignments.

### Utilize Potential Incentives as Needed to Fill Cross-Component Assignments

Our findings from Chapters Two, Three, and Four indicate that the military services, civilian government agencies, and the private sector have used various monetary and nonmonetary incentives to alter individuals' preferences in assignments. Different cross-component assignments may need different types of incentives. Therefore, as the services identify and expand the number of suitable cross-component positions, they should also think about the types of incentives (if any) that may be most appropriate for each cross-component assignment. These incentives are further elaborated on later in this chapter.

### Consider and Mitigate Effects of Changes on AC and RC Career Paths

Our findings in Chapters Three and Four also indicate that previous efforts to integrate resulted in some unanticipated or unintended consequences. For instance, when the Coast Guard integrated both its AC and RC under a single, unified AC chain of command, new opportunities for RC career advancement needed to be created because RC command positions were severely limited. However, the Coast Guard mitigated the loss of leadership positions because it believed the overall changes were needed and were beneficial for its total force moving forward.

One way to mitigate potential effects on both AC and RC career paths is to ensure that any changes do not impact adequate career development. For instance, some people we spoke with across the services also expressed concern that one of the unintended consequences of implementing cross-component command assignments may be that fewer RC command positions will be available to RC members because those positions will be designated as cross-component positions, ultimately hurting RC members' career advancement. Some of the individuals we spoke with suggested that RC members receive priority for RC command positions so that they can maintain career advancement opportunities in their RC.

Other potential ways to anticipate or mitigate the effects of potential changes are to create cross-component opportunities that are more flexible than longer-term traditional assignments. For instance, shorter cross-component assignment opportunities in the AC may make it easier for RC members to take those assignments. This could have major impacts on RC members' careers if cross-component assignments are rewarded or required for promotions, because RC members may not be able to apply for longer-term traditional assignments. Another way to mitigate the potential impacts of changes is to allow cross-component assignments to count as existing developmental or career-broadening requirements in lieu of creating a separate cross-component requirement for service members.

As was the case with Goldwater-Nichols, changes to assignment requirements such as those outlined above can ensure that the services will send their best individuals to AC/RC cross-component assignments. If promotion to senior rank requires an individual to have held an AC/RC cross-component assignment in lieu of a joint assignment, the services will automatically be incentivized to send their most qualified candidates to AC/RC cross-component assignments in order to make sure that they will be promoted.

## Changes to Promotion Processes

Like any incentive, changes to promotion processes could drive service member behavior regarding AC/RC integration. An excellent example of incentives driving individual behavior in the personnel management arena resulted from the enactment of Goldwater-Nichols. When DoD leadership began tracking promotion rates for joint officers, and when joint qualification became a requirement for promotion to general and flag officer rank, officers began to pursue joint assignments.

## Clarify Precepts and Board Charges

Unless civilian and military leaders are willing to take drastic steps and require specific promotion rates for specific groups of officers, military leadership can only "influence" promotion boards in limited ways. However, influence can have an impact.

For all of the services, officer promotion boards derive their authority from the same statute (10 U.S.C. 14308) and so they share the same basic framework. Members of promotion boards recommend officers to their service secretary for promotion based on a review and evaluation of the member's record and what the board believes is the member's ability to serve in the next higher grade. Although many boards "score" records on a predetermined numerical scale, officer promotion systems do not assign predetermined points for specific characteristics. As an example, officers do not accumulate a predetermined number of points for professional military education, overseas tours, duty in a combat zone, or any other part of their duty history. Unless the current promotion systems were completely redesigned, officers meeting promotion boards could not be given "extra points" for AC/RC cross-component assignments.

However, service secretaries can stress desirable characteristics and attributes and make sure each officer promotion board member evaluates a member's service in the context of service "priorities and special emphasis areas. . . ."[1] All promotion boards begin with direction from the service secretary to board members. This direction or guidance goes by different names (e.g., board precepts, board charge, memorandum of instruction). Regardless of the title, it is an opportunity for

---

[1]   Management-level reviews are part of the overall promotion process where critical promotion recommendations are finalized. Officers enter U.S. Air Force promotion boards with a Promotion Recommendation Form atop their record indicating "Definitely Promote," "Promote," or the rare "Do Not Promote" (Secretary of the Air Force Deborah James, "Memorandum of Instruction for Management Level Reviews," March 18, 2016).

service secretaries to tell board members what is important and what they should consider in their deliberations.

One common pitfall of board charges is guidance that is too broad or includes almost everyone and thus become ineffective. When everyone falls into a special category, no one is special. Board charges must be specific and not cover everyone. Over the years, board charges have been used to tell board members they should give special consideration to members who have served in a combat zone or members whose duties do not normally allow them to serve in a combat zone (e.g., missile launch officers in the Air Force). These instructions have also been used to emphasize the importance of acquisition, engineering, and language sills, to name a few.

One example of giving boards specific direction was seen when CJCS ADM Michael Mullen developed the Afghanistan/Pakistan (AFPAK) Hands program in 2009 and directed the services to inform promotion boards of the importance of the program and why officers participating in the program might have different career paths than their contemporaries. The AFPAK Hands program was modeled after a World War II program that focused on China. During World War II the term "China Hands" was used for American diplomats, journalists, and soldiers with deep knowledge of China or long and multiple tours in the area.

Using this approach, service secretaries could use the board precepts or charges to express the value of AC/RC integration. Rather than just commenting on the importance of AC/RC integration, guidance could be specifically focused on different programs considered important parts of AC/RC integration. Telling board members to consider command in one component equal to command in another is an example that would go a long way toward encouraging service members in the AC to accept command positions in the RC.

Policies similar to those put in place to track promotions after Goldwater-Nichols was enacted could also be put in place to track promotions for those who are serving in or have served in cross-component assignments, or to assign officers currently serving in cross-component assignments to promotion selection boards.

### Expand Board Membership

To further influence promotion boards, board membership could be better balanced to increase the emphasis on interoperability. By requiring a significant portion of promotion board membership from both AC and RC officers, the service secretary would be able to emphasize the importance of AC/RC integration and provide subject matter experts to the boards to answer questions about different jobs and level of importance of positions within their component. Increasing the number of AC officers who sit on RC promotion boards and the number of RC officers who sit on AC promotion boards would also foster greater appreciation for specific positions within each group.

However, it must be noted that such a change would also require additional training of board members, particularly related to differences in how AC and RC members fill out their evaluations. The different components tend to emphasize and reward different things in their evaluations, but this can be surmounted by educating board members about these differences.

### Develop an AC/RC Qualification System

This discussion highlights an important question: Is AC/RC integration critical enough to adopt a Goldwater-Nichols–type construct for promotions in general (below 0–7) or for advancement to general or flag officer specifically? Should all senior officers have experience serving in or with their component counterpart? If the answer to these questions is yes, a Goldwater-Nichols–type construct would be required.

Today, officers earn joint qualification through a specific process outlined earlier. Is that workable for AC/RC integration? Because of the relative size difference between the AC and the RC, it is unlikely that all officers could experience integration. However, those officers with the highest probability of advancing to senior positions could be broadened. By determining what might be meaningful integration (e.g., an assignment with the counterpart component or a deployment with the counterpart component), service assignment systems and the individual officers will begin to make AC/RC integration part of normal career development. There is one obvious caution.

Today, all services struggle with service requirements and joint requirements. The challenge of developing officers steeped in their service culture and operations and developing officers with broad joint experience is daunting. Services struggle today to get officers with enough service experience promoted within their service system and also with enough joint experience to be viable candidates for senior joint positions.

Goldwater-Nichols–type constructs have been established for other groups. In the Fiscal Year 1991 National Defense Authorization Act, Congress established Goldwater-Nichols–type promotion requirements and tracking for acquisition officers. As part of the Defense Acquisition Workforce Improvement Act (DAWIA), acquisition officers must be promoted at least at the same rate as other line officers in their service. DAWIA was the reaction to the 1986 President's Blue Ribbon Commission on Defense Management, also known as the Packard Commission. The report recommended "an alternate personnel management system" for a number of technically focused career fields (acquisition, contracting, scientists, and engineers) that would eventually be covered under the Defense Acquisition Corps umbrella.[2]

There is one major difference between joint qualified officers, acquisition officers, and officers experienced in both the AC and the RC. There is a finite requirement for acquisition officers in each service, and although the implementation of these acquisition requirements was well intended, it can overproduce acquisition officers.[3] This is not the case for officers steeped in both AC and RC experience.

As was the case with Goldwater-Nichols, the potential changes to promotion requirements listed above could ensure that the best individuals apply for AC/RC cross-component assignments. If there will be only a limited number of cross-component assignments and such assignments become a requirement or an advantage for promotion to

[2]  Albert A. Robbert, Tara L. Terry, Paul D. Emslie, and Michael Robbin, *Promotion Benchmarks for Senior Officers with Joint and Acquisition Service*, Santa Monica, Calif.: RAND Corporation, RR-1447-OSD, 2016.

[3]  Robbert et al., 2016.

senior rank, individuals will in turn be incentivized to seek out those cross-component assignments.

## Changes to Training and Education

### Expand Opportunities for Cross-Component Training

Another way to potentially facilitate cross-component integration is to expand opportunities for cross-component training and education so that AC and RC members interact with one another periodically throughout key parts of their careers. This strategy could also allow service members to interact with members from different components throughout their careers, and it may foster a holistic organizational culture and better understanding of the components.

As was the case with Goldwater-Nichols, such changes to training and education could also foster cross-component integration. By further emphasizing cross-component training and education, individuals will become increasingly familiar with the capabilities housed in the other components, as well as different cultures and processes across the components. This could not only facilitate better interoperability across services but also foster a better understanding of a broader DoD culture.

### Expand Opportunities for AC/RC Training Equivalence

While most training requirements are the same for service members from all components, there are exceptions. These exceptions tend to apply to those training and education opportunities that are lengthy, and therefore difficult for traditional RC service members to attend, and those that are costly in the context of limited resources. Resident intermediate-level education is an example of the former, where year-long resident education is limited almost exclusively to the AC and Active Guard Reserves. An example of the latter is the six-week CAP-STONE required of all newly promoted brigadier generals and rear admirals. The requirement is rigidly enforced for the AC, but only a few slots per class are set aside for the RC. While the reasons for de facto RC exclusion are understood, the fact that the exclusions exist contributes to different experiences and expectations for the AC and RC, thereby inhibiting integration. Questions for further study include the following: Is there greater benefit to furthering total force culture by ensuring equivalent training and education opportunities for both the AC and the RC? Or are there special categories of education and training in which the benefits to the AC outweigh efforts to provide equivalent training that may require shorter, less comprehensive, and less costly solutions to fully incorporate the RC? If so, what do such solutions look like?

## Incentives for Cross-Component Assignments

As indicated by our findings in Chapters Two, Three, and Four, both civilian and military organizations have used monetary and nonmonetary incentives to change individuals' preferences for assignments.

### Monetary
Our findings indicate that monetary incentives can be powerful in changing some people's assignment preferences. While the military cannot give service members the same types of monetary incentives that the private sector can offer its employees, the services could utilize several monetary incentives that they already employ in other contexts:

- assignment and incentive pays
- subsidies for housing[4]
- subsidies for childcare and family benefits.[5]

Different incentives could be used for different cross-component assignments, as well as to incentivize different types of individuals (e.g., individuals with families, single service members) to particular assignments. Assignment and incentive pays are already used by the services

---

[4] For instance, additional subsidies for housing could be used as an incentive in high-rent areas.

[5] For instance, subsidies for childcare and family benefits could be used as incentives if such services are not available at a nearby installation.

to incentivize service members to take particular assignments or enter particular occupations. Additional housing subsidies are already used by the services—particularly to augment standard housing allowances for all service members in expensive housing markets. Childcare and other family benefits could also provide incentives for service members with families to apply to cross-component assignments that offer these incentives. The services could publicize that these subsidies are available for locations that support cross-component assignments.

## Nonmonetary

One of the consistent themes in our findings in Chapters Two, Three, and Four is that while monetary incentives are powerful, nonmonetary incentives can be equally (if not more) powerful in enticing individuals' behaviors and preferences. Such incentives include the following:

- linking cross-component assignment to future assignments
- requiring cross-component assignments before selection for prestigious training and education opportunities
- offering work-life balance incentives
- providing award/ribbon/qualification for cross-component assignments.

We discuss each of these nonmonetary incentives below.

### Link Cross-Component Assignments to Future Assignments

One of the most persuasive incentives among employees of civilian agencies that we spoke with is the ability to link career-broadening or hardship assignments to future assignments. For instance, an assignment may become more desirable if applicants are given preference in their selection of follow-on assignments (e.g., they are allowed to apply to follow-on assignments earlier than other applicants, or they are given preference in follow-on assignment locations). Some of the military services are already using this type of incentive to fill certain assignments.

### Require Cross-Component Assignments Before Selection for Prestigious Training and Education Opportunities

The services could choose to require that service members serve in a cross-component assignment before they can be selected for prestigious training and education opportunities. Since these opportunities are highly sought in the military services, such an approach could provide a powerful incentive for individuals to seek out cross-component assignments.

### Offer Work-Life Balance Incentives

Individuals we spoke to from both the military and civilian government agencies indicated to us that work-life balance incentives are becoming increasingly salient to their service members and employees. Some of these incentives include the following:

- accommodating preferred assignment locations
- allowing flexible schedules, comp time, and vacation
- facilitating seamless transfer across components
- expanding and publicizing the CIP.

These are not new incentives used by the military, but they could be expanded to cross-component assignments. For instance, the military already uses assignment locations as an incentive for some positions, and it allows for some flexibility in work schedule under certain circumstances. The further expansion and increased publicizing of DoD's CIP is one way that DoD could help facilitate the seamless transfer across components. A senior Army leader told us that only seven people in the entire Army have utilized the CIP because soldiers do not know about the program.[6]

### Provide Award/Ribbon/Qualification for Cross-Component Assignments

There will always be a subset of individuals who are incentivized by tangible rewards such as awards, ribbons, or other qualifications.

In considering incentives for cross-component assignments, some lessons can be drawn from Goldwater-Nichols. Incentives at both the

---

[6]   Senior Army leader, discussion with the authors, February 24, 2017.

service level and the individual level can be automatically incorporated into changes to the assignment, promotion, and training and education requirements. Those changes can both incentivize the services to send their best candidates to cross-component assignments and incentivize the best individuals to pursue cross-component assignment opportunities.

## Potential Structural Changes

In addition to the changes in the previous four facets of our framework outlined above, strengthening AC/RC integration will also require potential foundational structural changes. These include potential changes to DoD strategy and doctrine, systems, and processes, as well as legal and regulatory changes. We discuss potential structural changes that could further facilitate AC/RC integration.

### Ensure Total Force Strategy Guides Integration Efforts

While Total Force policy is a ubiquitous term in DoD, Total Force strategy has not always guided the services' integration efforts. This is particularly important because the components' roles, responsibilities, force structures, and capabilities are all driven by the President's National Military Strategy, National Security Strategy, and Defense Strategic Guidance. However, as indicated in Chapter Two, the RC is not always included in strategic planning.

In its 2014 report to the Secretary of Defense, the Reserve Forces Policy Board (RFPB) reported that one of the factors inhibiting total force integration was the lack of an effective DoD Total Force policy. The RFPB report recommended that DoD should "develop and enforce a revised DoD Total Force Policy,"[7] and that "the services should better integrate their forces organizationally, in training, and during opera-

---

[7]  RFPB, *Reserve Component Use, Balance, Cost and Savings: A Response to Questions from the Secretary of Defense: Final Report to the Secretary of Defense*, RFPB Report FY14-02, Falls Church, Va., February 11, 2014, p. 13.

tional employment."[8] The report goes on to argue that "while the services have Total Force policies in place, the Department of Defense does not. This lack of total force perspective affects decision-making regarding the use of the RCs, AC-RC mix, and resourcing."[9] The personnel integration strategies explored in this study should be driven by a Total Force policy in which the RC is included in both DoD and service-level strategic planning.

## Implement Talent Management Workforce Strategy

The services currently have rigid human resources systems that include "one size fits all" paths for career progression. Some senior military leaders we spoke with indicated that, in order to achieve cross-component integration, the services will need to move toward an individual talent management workforce strategy that enables customized career paths and movement across components. The Army is forging ahead with this type of approach and has established a Talent Management Task Force that is leading an effort to manage human capital and individual skill sets.

## Address Legal/Regulatory Challenges and Undertake Efforts to Develop Mitigation Strategies

There is nothing in the law that explicitly forbids integrative efforts; instead, Congress explicitly contemplates the RCs as a trained and qualified force ready to supplement the ACs in a time of need (10 U.S.C. 10102; 32 U.S.C. 102). Efforts toward increased AC/RC integration will either have to work within the existing statutory framework in Title 10 and Title 32 or seek to revise and amend the existing law. These revisions or additions could help integration either by removing specific barriers or by implementing new provisions to undergird the integration efforts. These are two separate approaches, although they could be used in unison to provide for a smoother statutory environment that encourages greater integration of the components.

---

[8]   RFPB, 2014, p. 18.

[9]   RFPB, 2014, p. 13.

Assuming that no significant revisions to Title 10 or Title 32 are forthcoming, DoD already has significant latitude to promote greater AC/RC integration through the development of policy. As noted above, Congress has granted the power to administer the RCs to the secretaries of each service (10 U.S.C. 10202), in conjunction with the Assistant Secretary of Defense for Manpower and Reserve Affairs (10 U.S.C. 10201). Likewise, the RFPB (established in 10 U.S.C. 10301) is the independent adviser to the Secretary of Defense, charged with providing advice and recommendations on strategies, policies, and practices designed to improve and enhance the capabilities, efficiency, and effectiveness of the RCs.

These offices have the authority to issue regulation for the operation of the RCs and could design and support integration efforts of various scope, including training, equipping, promotions, and the like. Indeed, as discussed in Chapter Two, many services already have robust programs in place for closer AC/RC integration. That such efforts are under way—with significant successes—is evidence that there are existing authorities under which integration efforts can be made without statutory revisions. This is not to say that all obstacles are cleared; rather, the lack of explicit congressional inducements toward integration and statutory revisions is not sufficient to justify a lack of component integration efforts. In most cases, the services and OSD have the requisite authorities to drive integration if they choose to make this a priority.

While specific integration efforts might run afoul of specific procedural or substantive laws governing use of the RCs, the larger effort toward integration is not wholesale impeded by any existing statutory provision. For example, as NCFA pointed out, 10 U.S.C. 12304b actively contemplates AC/RC integration for some preplanned missions. But funding for those missions must be "specifically included and identified in the defense budget materials for the fiscal year or years in which such units are anticipated to be ordered to active duty."[10] Whether this is a legal hurdle is a matter of opinion; at a minimum, this provision sets a procedural stage that service personnel need to

---

[10] 10 U.S.C. 12304b.

anticipate in order to foster greater numbers of RC forces participating in preplanned missions.

In contrast to a service- or OSD-driven process for improving AC/RC integration, an alternative approach that utilizes statutory revisions as a driver of integration efforts could provide momentum and spur innovation in the ways active and reserve forces are integrated. For example, under current law, the service secretaries have the authority to order a reserve member to active duty, and the duties that member can undertake, include instructing and training active duty members of the armed forces (10 U.S.C. 12310). This approach leaves much to the service secretaries' discretion, including whether and to what extent RC members will be called on to train active duty personnel. Although the law contemplates the possibility of RC members training AC members, it does not require the services to establish a program whereby members of both the active and reserve forces conduct the training.

In response to this situation in which the services have such significant discretion over integration efforts, Congress could require more robust integration programs (either as a pilot program or on a larger scale). Statutory efforts, such as pilot programs in DoD annual authorizations, could be a useful way to give congressional consent and encouragement to efforts that experiment with increased integration of active and reserve forces. They could, for example, require more integrated training and education, increase the maximum number of (and establish minimums for) RC personnel in preplanned missions, and increase the presence of RC members at service headquarters. Whether congressional action in these affairs is desirable is subject to much debate. The point here is that the legal barriers to integration are changeable.

A statutory approach to requiring the implementation of organizational change could, in this instance, be overwrought: assuming the services (and DoD more broadly) have sufficient current authorities to encourage integration, statutory reform through Title 10 and Title 32 would be a disproportionate involvement by Congress in DoD's planning and force management processes. Congress could, alternatively, in its annual authorizations require pilot programming that directs the service secretaries to design and evaluate specific integration programs (perhaps in conjunction with the Assistant Secretary for

Manpower and Reserve Affairs and/or the RFPB). Given the differences across the services in their missions, cultures, and the degree to which the active and reserve forces are already being integrated, a flexible approach would have a higher likelihood of producing integration policies that meet the needs of the services individually.

### Duty Status Reform

In addition to the broader statutory issues mentioned above, specific statutory changes could also facilitate AC/RC integration. For instance, there are currently about 30 different duty statuses that RC members switch between, depending on the type of orders they are on, how their assignment is funded, and the mission they are performing. As RC members switch from one duty status to another, their pay and benefits may also change. This complicated system can cause barriers to cross-component integration. DoD is currently undergoing a review of the current duty status system and is considering simplifying it. Such reform could facilitate AC/RC integration more generally and could simplify the implementation of cross-component assignments more specifically.

### Improve Scrolling Process

In addition, during our discussions with senior leaders of all the services, we learned that "scrolling" (the transfer of a service member from one component to another) is currently a time-consuming process that in some cases can take six to seven months. The problem is particularly acute in the Air Force;[11] however, we heard about this issue from the other services as well. Streamlining the scrolling process could also facilitate greater movement across components.

### System Changes

In addition to the structural changes discussed above, DoD can also make changes to its systems to facilitate AC/RC integration. This includes continuing to implement IPPSs, which allow the services to use

---

[11] Senior Air Force leader, discussion with the authors, March 1, 2017; Senior Air Force leaders, discussion with the authors, November 22, 2016.

a single personnel management database for both active and reserve component members.

### Fully Implement IPPS

While the Coast Guard, Marine Corps, and Navy have rather robust IPPSs, the Air Force and the Army are still working to complete their systems. An Air Force participant in our discussions emphasized that systems remain a barrier to integration in the Air Force:

> I think Continuum of Service is a really important issue that we have made some great strides on. But in order to have more seamless integration, our systems need to be able to facilitate this. For example, the personnel and finance systems must be updated to ensure that integration is seamless.[12]

During our informational discussions, senior members of the Coast Guard, Marine Corps, and Navy agreed that IPPS was key in facilitating permeability across components. Continuing to fully implement IPPS across all of the services is a structural change that could go a long way toward increasing AC/RC integration, facilitating cross-component assignments, and enabling greater permeability across components. IPPS can also facilitate readiness because it allows commanders to identify skill sets across the components.

## Considerations Regarding When to Offer Cross-Component Assignments

As the services think about ways to incentivize service members to participate in cross-component assignments, when those assignments are offered in a person's career may be particularly important. If cross-component assignments are viewed as an extra requirement or a burden to an individual's career progression, he or she may be deterred from applying for such an assignment. However, if cross-component assignments are viewed as career enhancing (or at least not detrimental to

---

[12] Air Force leaders, roundtable discussion with the authors, February 17, 2017.

one's career), individuals may be incentivized to participate in such assignments. Our literature review and our discussions with individuals from both civilian and military organizations revealed that there are two primary perspectives on when career-broadening programs should be offered to individuals: (1) starting very early on and throughout a person's career, and (2) from midcareer onward. We explore the pros and cons of each of these approaches below.

Many of the people we spoke with in both military and civilian organizations indicated that one way to develop a holistic organizational culture is to emphasize integration from the time a person begins working in an organization. Doing so can expose an individual to all of the different components of an organization and give him or her a better sense of the capabilities of each component. This emphasis on integration as a priority can then be reinforced periodically throughout an individual's career.

Other people we spoke with thought that the optimal time to begin to offer cross-component assignments is from the middle of an individual's career onward. We found that many civilian organizations do not offer career-broadening experiences to their employees until later in their careers because they want junior-level employees to focus on developing core competencies before they broaden out.[13] Most military representatives we spoke with said that they thought the optimal time to begin to offer cross-component assignments is at the junior officer level. They indicated that it was important for military recruits and trainees to focus on developing core competencies early on in their career before they expanded their focus to career-broadening experiences.

Regardless of when cross-component assignments are first offered in an individual's career, opportunity points can be identified where cross-component assignments can become part of typical career paths and progression. The addition of another requirement could be a major disincentive for some individuals to participate in cross-component assignments; therefore, by deliberately ensuring that cross-component assignments do not add an additional burden, individuals could be

---

[13] We found that this is the case for career-broadening programs at the State Department, DIA, and OUSDP.

incentivized to participate in them. When such opportunities are identified throughout an individual's career, they can also serve to continually reinforce a holistic organizational culture.

In developing its pilot TFAP program, in which junior AC captains take a two-year assignment in a USAR unit, the Army thought deliberately about when it could offer cross-component assignments to AC officers at a point in their careers when such an assignment would not have a negative impact on their career progression. The Army identified that the ideal time to offer cross-component assignments is right after junior officers have graduated from the advanced course. During this time, these officers are primarily doing staff work; therefore, a two-year cross-component assignment offers them considerable opportunities (including an opportunity to command a unit) at a time in their careers when they otherwise would not have many opportunities.

# CHAPTER SIX

# Conclusion

Our findings indicate that it is possible to enhance cross-component knowledge and awareness and further development of a total force culture. However, as indicated in Figure 5.2, additional steps need to be taken to modify and align personnel policies to achieve these objectives. Most importantly, changes to assignment and promotion policies are critical for accomplishing these objectives. The historical analogue of DoD joint integration highlights the importance of assignment and promotion policies in aligning personnel policies to DoD priorities. Our findings from the practices used by the services, foreign militaries, other U.S. government agencies, and the private sector also indicate that various incentives (monetary and nonmonetary) can facilitate cross-component integration.

## What DoD and the Services Should Do Next

Senior DoD and service leadership can take several steps to help facilitate deeper AC/RC integration. First, they should decide whether integration is a priority. If so, top service leadership should clarify the purpose of integration and establish goals and benchmarks for furthering integration. Next, the services should undertake a review of current assignment and promotion policies and determine how they will change. This includes (1) identifying positions for cross-component assignments, (2) determining the number and grade levels of assignees, (3) implementing a program of incentives as needed, and (4) altering promotion policies and practices to reward cross-component service.

Last, DoD should continue to make progress on structural changes such as addressing legal/regulatory barriers to integration and implementing system changes that could facilitate AC/RC integration. Such actions would set the foundation for any further efforts to integrate the active and reserve components and to achieve the development of a stronger total force culture. Without such a starting point, any follow-on efforts will likely be disjointed and ineffective.

In addition, our findings indicate that there are several overarching steps that the services and DoD could take to facilitate implementation of any efforts to deepen AC/RC integration: (1) define the purpose of AC/RC integration efforts, (2) foster a shift in culture and leadership buy-in, (3) tailor integration efforts to unique service force structure and RC competencies, and (4) evaluate integration initiatives.

### Define the Purpose of AC/RC Integration Efforts

Among the cautionary tales that our analysis of the literature and our discussions with both senior service leaders and representatives from other government agencies provide is that cross-component integration efforts are likely to fail if the purpose of integration efforts is not clearly defined by leadership. During our discussions with service representatives, the question of why further AC/RC integration is needed came up repeatedly (especially among senior officers). For instance, an Air Force leader we spoke with cautioned that "integration solely for the sake of integration is not the right reason to integrate. Integration can be synergistic, but we have to make sure we don't kill a unit we are trying to integrate."[1] If DoD and the services plan to make further AC/RC integration a priority, they will need to explain why such integration is needed, and what the business imperative or value-add of such integration is. That message will then need to be communicated down the chain of command.

### Foster a Shift in Culture and Leadership Buy-In

In our discussions with both senior leaders in the services and representatives from other government agencies, there was consensus that

---

[1]   Air Force leaders, roundtable discussion with the authors, February 17, 2017.

increasing integration across the active and reserve components would require a shift in culture and support from DoD and service leadership. Several individuals also indicated that previous attempts at integration in the services have been dependent on the level of buy-in from leadership, and that the pace and degree of integration were dependent on the level of leadership support for those efforts.

Leadership (at all levels of the chain of command) is also key to setting the tone for the integration process. Moving forward, if DoD plans to continue to make AC/RC integration a priority, that message should be clearly communicated from senior DoD and service leaders. Without commitment from key stakeholders and without visible involvement by senior leaders, progress on integration is difficult or impossible to achieve. Integration also needs to be supported by legal and policy changes, and senior leaders are uniquely positioned to implement and enforce these types of changes.

## Tailor Integration Efforts to Unique Service Force Structures and RC Competencies

Our discussions with the services clearly identified the different approaches they have taken to AC/RC integration. These approaches have been driven by the services' missions, force structures, and competencies that reside in their RCs. As DoD contemplates different strategies for deepening AC/RC integration, these underlying reasons for the services' existing integration approaches are important to keep in mind. While a "one size fits all," top-down mandated approach such as Goldwater-Nichols may be a means by which AC/RC integration is made mandatory, it may also have unintended consequences across the services. For instance, such an approach could have minimal impacts on the Army (whose RCs are equipped and manned in many ways as mirror images of its AC), but it could have far-reaching impacts on the Navy (whose RC is equipped and manned differently than its AC, and therefore may cause extensive force structure changes). Therefore, as DoD moves forward in deepening AC/RC integration, the services will need some flexibility in how they implement any AC/RC integration priorities. Such flexibility will allow them to develop the AC/RC integration initiatives that are most appropriate to each service, given their different missions, force structures, and RC competencies, and how they utilize their RCs.

### Evaluate Integration Initiatives

Our findings from our discussions with senior DoD and service leaders also indicated the importance of evaluating integration efforts so that they can be improved. We could not find any instances in which previous DoD and service efforts at improving AC/RC integration had been formally evaluated. In order to be able to evaluate integration initiatives, it will be critical for DoD and the services to clearly define the purpose and desired end state for AC/RC integration so that integration initiatives can be evaluated and measured against those objectives. As DoD and the services implement new AC/RC integration initiatives, it will be important to evaluate them in order to make adjustments along the way and improve their outcomes. It will also be important to track the participants of these initiatives in order to identify the impact that those integration initiatives had on their career progression.

## Closing Thoughts

Our research identified many models both within the services and in other organizations that are designed to enhance cross-component knowledge and awareness and further development of a total force culture. These models include rotational programs, various monetary and nonmonetary incentives used to encourage cross-component experience, and changes to assignment and promotion processes. As DoD and the services consider ways to deepen AC/RC integration, these models and the lessons learned from other integration efforts could serve as important guideposts.

# Case Studies of U.S. Government Civilian Programs

Through our research, we identified three civilian agencies whose integration efforts provide insights applicable to the U.S. military's efforts to move toward a total force. Incentive structures, challenges, and lessons learned from each of these programs are analyzed below.

## DIA

DIA manages its rotational programs as part of its Talent Management initiative, called Rank in Person. Rank in Person is composed of three processes: career assignments, promotions, and career development. Under the Rank in Person system there are ten career fields (such as analysis, human intelligence, and information technology), each of which has a specified career path guide that defines the competencies each career field is required to master. These career fields are intended to create a common culture among job series that have similar competencies and responsibilities.

Rank in Person is designed to foster professional development through a standardized process that emphasizes both agency-wide and career-specific skills. DIA has pursued the Rank in Person program "to build a more robust and agile workforce that provides intelligence on foreign militaries and operating environments that delivers decision advantage to prevent and decisively win wars; all while adapting to uncertain changes to the strategic, operational, and fiscal environment."[1] Through

---

[1] Vincent R. Steward, quoted in DIA, "DIA's Talent Management System Overview," briefing slides, Washington, D.C., 2016, p. 3.

the Rank in Person system, DIA seeks to increase DIA officer professionalism and expertise in both specialized and common skills, build a common workforce culture based on continuous learning, and develop a standardized promotion system that rewards career development.[2]

People we spoke with from DIA further explained that the Rank in Person system is designed to "ensure our officers have breadth and depth, and that skill sets are transferrable."[3] This mission is driven in part by DIA's effort to ensure that its employees can flexibly fill several roles as the mission changes. DIA endeavors to retain its employees long term, using the motto "From seeds to trees" to highlight its attempt to grow analysts from junior levels to senior management positions.

Within Rank in Person is the Annual Career Assignment Program (ACAP). ACAP applies to employees starting at the GS-13 level and encourages them to compete for lateral positions in the United States and overseas. ACAP does not apply to more junior employees, because DIA directs those individuals to develop core competencies before expanding their skill set beyond their core career field. The positions located overseas are accompanied by specific incentives, largely financial. Some of these incentives include nontaxable living quarters allowance, post allowance, increased annual leave, and an additional retirement annuity benefit equal to 1.7 percent for each year served overseas.[4] Additional specific incentives do not accompany hardship assignments, since most of those assignments have enough volunteers. DIA officials relayed that these assignments are generally filled due to a combination of factors, including financial incentives and recruitment of employees who desire to gain deployment experience or provide direct support to U.S. operations overseas.

Part of DIA's rotational programs are details to military senior service schools, such as the Naval War College in Newport, Rhode Island. Currently, there is no standardization for how employees are chosen for senior service schools, but DIA anticipates developing cri-

---

[2]   DIA, "DIA's Talent Management System Overview," 2016, p. 5.

[3]   DIA representatives, discussion with the authors, January 30, 2017.

[4]   DIA, "Fiscal Year 2017 Annual Career Assignment Program (ACAP) Way Forward," briefing slides, Washington, D.C.: DIA Office of Human Resources, April 8, 2016.

teria in the near future. Finally, when a DIA employee pursues a JDA, a cross-agency assignment, or an overseas deployment, he or she will eventually return to his or her original position upon completion of the temporary assignment.

ACAP is closely aligned with DIA's promotion process. Promotions are based on a point system, where employees must accumulate at least 80 points to be promoted. However, the total number of employees promoted every year is largely dependent on how many billets are available. For promotion consideration from GS-13 to GS-14 and from GS-14 to GS-15, employees are given three points for cross-agency assignments (such as from management to analysis), deployments in support of DIA, and JDAs to another federal agency, for a maximum of nine points. Rotations conducted in the five prior years can count toward promotion points.

Rather than requiring these rotations, DIA strongly encourages them. DIA seeks to allow for variation to evaluate individuals on a case-by-case basis, and to not unfairly disadvantage certain employees if a rotation is not feasible with their lifestyle or career at a given time.

### Challenges and Ways to Overcome Them
Under the DIA system, if an employee is eligible for a GS-14 or GS-15 promotion, he or she must identify a billet that is available at that grade. If the only billet happens to be in another location (for instance, at a combatant command overseas) and that employee is not able to relocate, the employee cannot receive a promotion and stay in his or her current billet. This prevents otherwise eligible employees from applying for promotions, because if an employee does receive a promotion, he or she is not allowed to stay within his or her current billet because the number of employees at a specific grade is tightly controlled.

DIA representatives also described a mismatch between incentives and the employees they target. The promotion incentives are targeted at GS-13s and GS-14s, who are more likely to have families and unable to be as mobile as the more junior employees at DIA who are not offered incentives to take rotational assignments outside their home area.

### Lessons Learned

ACAP is too new to be fully analyzed, but the people we spoke with from DIA told us that it appears as if more employees are eager to pursue JDAs. Prior to this change, JDAs were not rewarded, and supervisors were hesitant to support rotations because they were concerned about losing an employee for a year or more. According to one DIA representative from human resources, the culture at DIA is changing with the junior employees, who tend to favor broadening assignments and are recruited in part due to their willingness to work in overseas environments.[5]

### DoS—Foreign Service Program

DoS's Foreign Service Officers (FSOs) compose the bulk of the Foreign Service and are responsible for formulating and executing the U.S. government's foreign policy objectives. Generalist FSOs are required to be proficient in a number of DoS functions. DoS employs several types of rotation-based programs. The primary structure that governs generalist Foreign Service assignments is fundamentally rotational: FSOs are assigned to posts or other positions and then move to another post generally two to three years later. This structure is intended to create experienced diplomats, ease the burden of challenging overseas assignments, and fill critical needs roles for DoS worldwide. The assignment process is closely linked to DoS's promotion process for its FSOs, which also encourages broadening assignments within DoS, and includes long-term training and detail assignments outside DoS as well.

### Career Assignments

State FSOs fill positions at posts globally, which include high-demand positions in places such as Western Europe and Australia, and positions that are more challenging to fill either because of danger or living conditions or because of a lack of personnel with required skills available at a certain time.

To fill its positions worldwide, DoS employs an annual prioritized bidding process. First, DoS prioritizes assignments to Priority Service

---

[5]  DIA human resources representatives, discussion with the authors, January 30, 2017.

Posts (PSPs), which include certain countries where active armed conflict is occurring and threats to U.S. officials are high. Once those assignments are filled, DoS advertises training and detail positions, and opens the general assignment bidding process for all other remaining jobs, both overseas and in the United States.

Beyond the PSP, DoS also has several other posts that are considered hard-to-fill positions. Some of these posts are overseas, and some are located in Washington, D.C. A position is determined to be hard to fill if fewer than three minimally qualified FSOs apply to the position at the time DoS reviews the qualifications of the candidates. When a position is flagged as hard to fill, DoS can offer early preference to FSOs who are "stretching," which is when an individual can bid on a position beyond his or her own pay grade.

To incentivize its FSOs to take PSPs or other hard-to-fill assignments, DoS employs several mechanisms. First, DoS offers an incentive called "linking," which allows the FSO to bid a year earlier than other FSOs for the assignment that will follow the PSP. Only FSOs who are undertaking PSPs can compete for the linked positions in that time frame, substantially narrowing the pool of competitors for highly desirable positions. DoS also allows the FSOs' dependents to remain overseas while the FSO is in the PSP location, a practice that is not supported under other circumstances because of the cost of housing dependents overseas. These incentives are in addition to the increased pay offered in these assignments, such as hardship pay (15 percent or more of base pay) and differential/danger pay (an additional percentage of base pay). DoS also allows PSPs to "stretch" into a position one pay grade above theirs for their linked assignment. Finally, DoS offers additional paid rest and recuperation travel. According to DoS personnel management officials, the strongest motivator for its employees to bid on PSP or hard-to-fill or critical positions is their role in the promotion process, described in the next section.

DoS also uses a similar linking model to fill some of its nonhardship but hard-to-fill positions. For instance, congressional fellowships that DoS commits to annually can be challenging to fill, so they are linked to follow-on assignments the way that PSPs are linked. By volunteering for congressional fellowships, applicants are able to receive preference of consideration earlier than their peers for the following year's assignments.

For its hard-to-fill positions with fewer than three applicants, DoS allows employees to bid early and "stretch" to fill positions one pay grade above their current band. Further, if an employee fills a hard-to-fill position for three years instead of the traditional two years, State will increase the FSO's salary by 15 percent. The FSO must complete the full three years in order to receive the 15 percent, to prevent the individual from accepting the incentive pay and leaving the post after two years.

DoS also supports long-term training or detail assignments, which are used both as broadening assignments for its employees and, on occasion, as incentives for employees who take challenging assignments. While not the only criterion considered in selecting an applicant for a detail or educational opportunity, people we spoke with from DoS did note that such experience could increase an applicant's chances of selection for competitive opportunities. In the civil service, if a DoS employee takes a long-term training assignment, that individual returns to his or her original position. However, in the rotation-based FSO community, once an employee takes a detail assignment, he or she does not return to a particular office, but rather moves into the next position that the FSO has been placed in through the bidding process.

### Promotion Process

In order to promote in the Foreign Service system, FSOs must meet the career development requirements specified earlier. According to the DoS representatives with whom RAND spoke, emphasis is increasingly placed on outside experience as well. Some of these outside experience assignments include long-term training (such as U.S. military war colleges and academic institutions such as Princeton and Tufts) or detail assignments to nongovernment organizations, U.S. Congress, or the National Security Council, and exchanges with other government agencies such as DoD or U.S. Agency for International Development. These details are pursued through a competitive process run by DoS and are usually two years or more in duration.

The FSO's career development program is currently undergoing revision, but as of January 2017, the requirements for generalists included the following:[6]

1.  operational effectiveness, which includes a "major" and a "minor" in regional assignments
2.  leadership effectiveness, which includes leadership and management training at each grade
3.  language proficiency in at least one language at the 3/3 level
4.  service needs, which includes service at an assignment whose hardship rating is 15 percent or greater, a post that incurs danger pay, or two directed entry-level tours.

Additionally, Foreign Service generalists are assigned to one of five career "cones": political, economic, consular, management, and public diplomacy. Foreign Service generalists are strongly encouraged to pursue "out of cone" assignments, which tend to be a year in length, in order to be promoted to senior levels.

For instance, to be considered for selection as a deputy chief of mission or another management post, an individual will generally need to demonstrate broad experience among the cones, as that individual will be expected to employ skills drawing from several of the cones. However, DoS does not specifically require this type of experience, which allows the organization to consider individuals' specific experience for each position.

Selection boards may use difficulty of prior assignments as a criterion to determine whether an FSO should be promoted. While only one factor among a range of others, succeeding in a challenging position may cause the selection board to view that candidate more favorably compared with another who completed a more routine assignment.

### Challenges and Ways to Overcome Them

DoS's incentive to link PSP or certain hard-to-fill assignments to early consideration assignments for the following year has drawbacks,

---

[6] "Career Development Program Requirements for Foreign Service Generalists," handout from U.S. Department of State, undated.

according to personnel managers at DoS. The posts receiving the linked personnel must select among the early applicants if they are minimally qualified, which reduces the number of personnel that they may choose from, and may force the post to employ an FSO that they otherwise would not have selected.

### Lessons Learned

Many of the rules within the Foreign Service Career Development Program are not enforced, but are considered guidelines. This is also true in certain civil service positions, such as Diplomatic Security.[7] This allows the selection board flexibility to weigh certain factors, such as several overseas tours, even if other criteria have not been met. According to one diplomatic security agent, despite the inconsistency that the guidelines pose, their flexible nature overall benefits DoS in allowing individuals to be considered on a case-by-case basis.[8]

### OUSDP

OUSDP, whose primary mission is to "consistently provide responsive, forward-thinking, and insightful policy advice and support to the Secretary of Defense, and the Department of Defense, in alignment with national security objectives,"[9] employs multiple initiatives aimed at achieving greater integration of its employees. Generally, OUSDP's integration efforts are concentrated in its promotion system, its external development assignments, and, recently, its development of a formal program to introduce new hires to OUSDP as a cadre. Overall, OUSDP uses two incentives to promote its personnel to work in hardship positions: tools that enable work-life balance, such as alternative work schedules, and long-term training assignments upon completion of a hardship assignment.

Most OUSDP staff officers fill roles in regional offices (e.g., managing a bilateral relationship on a country desk), functional offices (e.g., cyber policy or force development policy), or support and management

---

[7]   Member of the Diplomatic Service, discussion with the authors, September 27, 2016.

[8]   Member of the Diplomatic Service, discussion with the authors, September 27, 2016.

[9]   Office of the Secretary of Defense for Policy, home page.

offices. Both the regional and the functional offices tend to have different leadership styles, skill sets, and cultures. Staff officers may remain in a given role indefinitely, but OUSDP personnel are encouraged to gain experience by applying for a new position every few years—leaving the prior position vacant and open for applicants when moving to a new role. Overall, the OUSDP representatives we spoke with said that they believe the rotational employment model with an emphasis on both regional and functional experience has been helpful in inculcating a holistic OUSDP culture. Rotations among various offices can create larger professional networks and greater understanding of the various roles and functions OUSDP plays.

### Promotion Process

Through its promotion process at the GS-14 and GS-15 levels, OUSDP emphasizes multidisciplinary experience. Overseen by a panel of OUSDP employees called the Career Development Board (CDB), the organization requires mastery of four skill categories (international, intergovernment, inter-DoD, and leadership) for promotion to GS-15, and three of the four for GS-14 promotion. Applicants are required to obtain a nomination from their home office before submitting an application to the nominations board—a competitive endeavor that ultimately results in most applicants that appear before the CDB gaining promotion. However, in order to be nominated by a home office and ultimately recommended for promotion (particularly to GS-15), an applicant must have experience in both regional offices and functional offices.

OUSDP does not include difficulty of a staff position in its promotion deliberations, emphasizing that not all of its employees will flourish in a hardship position, and OUSDP does not want to disadvantage an employee that is excelling in a role that remains critical—such as managing a bilateral relationship with a major defense partner—but that is not hard to fill.

### Hard-to-Fill Positions

OUSDP has focused additional attention on hard-to-fill positions, such as human resources–related positions and high-stress portfolios such as director of the Russia or Syria portfolio. In the past, OUSDP avoided directed assignments, but in 2016 many critical positions were vacant.

OUSDP conducted a survey among its personnel and asked what incentives would compel them to pursue a hardship assignment. Through this survey, OUSDP gained valuable insights: first, that many employees would simply take a hardship job if leadership asked; second, that paid time off and comp time were more valuable than additional pay; third, that the type of position employees were willing to take depended substantially on demands generated from their personal life, such as caring for young children or an aging parent; and fourth, that there was concern about what specific responsibilities the employee would have, as personnel were hesitant to leave their current job to take on a role that might entail undesirable work. As a result, OUSDP designed a contract in which the employee taking the hardship assignment would choose their incentive—for example, compensation time or a compressed work schedule. However, this contract mechanism has not been implemented widely, as OUSDP was able to bring on several David L. Boren fellows at the same time to fill personnel needs.

Given the hiring and resource constraints that have challenged OSD in recent years, OUSDP has been forced to creatively approach filling vacant billets. In addition to Boren fellows, OUSDP uses no-cost JDAs from the IC and Intergovernmental Personnel Act Assignments to populate its organization with policy action officers from different organizations. These detailees serve in critical positions, taking on responsibilities equal to those of permanent party staff in OUSDP.

### Long-Term Developmental Assignments

Another area in which OUSDP promotes integration is with its sponsorship of long-term developmental assignments. In these assignments, a staff officer will apply for a detail position to an external agency such as the National Security Council, an educational opportunity such as a year at the National War College, or an exchange to the United Kingdom's Ministry of Defence. OUSDP supports these assignments, as they provide broadening skills to its employees, improve employee morale, and increase its staff officers' familiarity with and networks in the various organizations with which OUSDP collaborates. Long-term developmental assignments are not used as a factor for promotion in OUSDP, but rather are treated as a reward in some instances for employees that have been working in a particularly stressful position. Conversely, if an employee has taken advantage of multiple long-term developmental assignments, his or her promotion application may be scrutinized more carefully, as service to OUSDP in a promotion period is greatly valued.

### Fostering a Common Organizational Culture

In attempting to foster integration among new hires, the Human Capital directorate in OUSDP recently launched a new onboarding program designed to teach OUSDP history, administrative details, and values and to foster a sense of community throughout the organization. Composed of new action officers throughout OUSDP, the program is intended to inculcate cohesion among its new hires that could theoretically build a common culture across OUSDP.

### Challenges and Ways to Overcome Them

Integration remains difficult to achieve among certain components within OUSDP. Success in OUSDP is heavily reliant on one's ability to develop and utilize networks throughout the OSD organization and with OSD's counterparts in the Joint Staff and the services, and throughout the interagency. However, some offices have insular leadership, and the staff tend to reflect their leadership's personality, focusing internally and refraining from collaboration with other OUSDP entities that would facilitate greater cultural development.

If an OUSDP employee pursues a long-term training or developmental opportunity outside the organization, such as a detail to another U.S. government agency, the position within the specific office in OUSDP opens to new applicants. Upon return from the detailed assignment, the employee must find an open position within OUSDP, but is not guaranteed a particular role or office. This lack of guaranteed placement is challenging for OUSDP and likely deters certain employees from pursuing developmental assignments outside the organization. At present, OUSDP has not forged a solution to this challenge, but applications to long-term developmental assignments remain high.

OUSDP also feels that it does not sufficiently advocate for the individuals who willingly fill hardship positions. The CDB attempts to evaluate each individual on his or her own merit, understanding that OUSDP's most talented employees and best potential leaders are not necessarily those with the lifestyle and temperament most suited to hardship positions.

While OUSDP employees are not awarded or penalized for taking hardship positions, they may receive additional scrutiny from the CDB if they have spent time pursuing long-term developmental assignments outside the OUSDP organization. According to one CDB representative, the question of "what have you done for us lately?" is raised when considering these applicants for promotion.[10]

### Lessons Learned

Requiring the rotations poses substantial challenges. One staff officer may thrive in a lower-intensity but still critical environment within OUSDP, such as managing the bilateral relationship with a friendly country, but perform poorly in a hardship assignment that requires different skills and stress tolerance. OUSDP prefers to utilize its personnel to their own maximum potential, and would view required rotations as punishing those who thrive in more traditional environments.

The people we spoke with from OUSDP did not feel that applicants who elect to fill hardship positions should be rewarded in the promotion process, noting that if an employee merits a promotion, his or her potential within OUSDP will already be well known and will not necessarily be reliant on that hardship position.

They noted that, especially because its employees tend not to be motivated by money, it tries to base its approach to talent management on Dan Pink's concept of Mastery, Autonomy, and Purpose, which explains that these factors create intrinsic motivation that outweighs extrinsic rewards.[11] They noted, "We found if people could guarantee mastery of some kind, and autonomy of some kind, and purpose of some kind, then people would stay."[12]

---

[10]  OUSDP officials, discussion with the authors, January 23, 2017.

[11]  D. H. Pink, *Drive: The Surprising Truth About What Motivates Us*, New York: Riverhead Books, 2009.

[12]  OUSDP officials, discussion with the authors, January 23, 2017.

They also recommended collaborative projects for AC and RC members to work on together, based on the daily tasks that require cross-directorate coordination within OUSDP. In instances where details or direct assignments between the two components are not feasible, the appropriate service could assign a specific project—for instance, a task force to develop Total Force legislative proposals for an upcoming authorization or appropriations bill—that requires equal contribution from both AC and RC members of the same unit, or from different AC and RC units.

# Abbreviations

| | |
|---|---|
| AC | active component |
| ACAP | Annual Career Assignment Program |
| AFPM | Air Force Policy Memorandum |
| ANG | Air National Guard |
| ARNG | Army National Guard |
| AUPP | Associated Unit Pilot Program |
| CDB | Career Development Board |
| CIP | Career Intermission Program |
| CJCS | chairman of the Joint Chiefs of Staff |
| CJCSI | chairman of the Joint Chiefs of Staff Instruction |
| DIA | Defense Intelligence Agency |
| DoD | Department of Defense |
| DoS | Department of State |
| FSO | Foreign Service Officer |
| HS | "have served" joint assignments |
| I&I | Inspector and Instructor |
| IC | Intelligence Community |
| IMA | Individual Mobilization Augmentee |

| | |
|---|---|
| IPPS | Integrated Pay and Personnel System |
| JCS | Joint Chiefs of Staff |
| JDA | Joint Duty Assignment |
| JDP | Joint Duty Program |
| JQO | Joint Qualified Officer |
| JTF | Joint Task Force |
| MCU | multicomponent unit |
| NCFA | National Commission on the Future of the Army |
| NCO | noncommissioned officer |
| NCSAF | National Commission on the Structure of the Air Force |
| ODNI | Office of the Director of National Intelligence |
| OSD | Office of the Secretary of Defense |
| OUSDP | Office of the Under Secretary of Defense (Policy) |
| PSP | Priority Service Post |
| RC | reserve component |
| RFPB | Reserve Forces Policy Board |
| SI | "serving in" joint assignments |
| TFAP | Total Force Assignment Program |
| USAR | United States Army Reserve |
| U.S.C. | United States Code |
| USCG | United States Coast Guard |
| USMC | United States Marine Corps |

# References

"About Fighting and Winning Wars," an interview with Dick Cheney, *Proceedings*, May 1996, p. 32.

Air National Guard Instruction 36-101, *Air National Guard Active Guard/Reserve (AGR) Program*, June 3, 2010.

Air National Guard Instruction 10-203, *Air National Guard Alert Resource Management*, February 22, 2012.

Amburn, Brad, "The Unbearable Jointness of Being," *Foreign Policy.com*, November 16, 2009. As of December 18, 2018:
https://foreignpolicy.com/2009/11/16/the-unbearable-jointness-of-being/

Anderson, David O., and J. A. Winnefeld, "Navy's Reserve Will Be Integrated with Active Forces," *Proceedings*, September 2004, p. 61.

ANGI—*See* Air National Guard Instruction.

Australian Government, Department of Defence, "ADF Total Workforce Model." As of January 10, 2018:
http://www.defence.gov.au/ADF-TotalWorkforceModel/

Bailey, Kat, Air Force Personnel Center Public Affairs, "AF Adds International Affairs to VLPAD Program," San Antonio–Randolph, Tex., June 2, 2017.

Barki, Henri, and Alain Pinsonneault, "A Model of Organizational Integration, Implementation Effort, and Performance," *Organization Science*, Vol. 16, No. 2, 2005, pp. 165–179.

Beckhard, R., and R. Harris, *Organizational Transitions: Managing Complex Change*, 2nd ed., Reading, Mass.: Addison-Wesley, 1987.

Bergman, John W., "Marine Forces Reserve in Transition," *Joint Force Quarterly*, No. 43, 2006, pp. 26–28.

Biljsma-Frankem, Katinka, "On Managing Cultural Integration and Cultural Change in Mergers and Acquisitions," *Journal of European Industrial Training*, Vol. 25, Nos. 2/3/4, 2001, pp. 192–207.

Boeing, "Rotational Programs," *Boeing.* As of January 10, 2018:
http://www.boeing.com/careers/college/rotational-programs.page

Brinkerhoff, John R., and Stanley A. Horowitz, *Active-Reserve Integration in the Coast Guard*, Alexandria, Va.: Institute for Defense Analyses, 1996.

Brown, Graham, Thomas B. Lawrence, and Sandra L. Robinson, "Territoriality in Organizations," *Academy of Management Review*, Vol. 30, No. 3, 2005, pp. 577–594.

Burke, David, "The Surprising Benefit of Work Sabbaticals," *Forbes*, June 29, 2016. As of April 28, 2020:
https://www.forbes.com/sites/davidburkus/2016/06/29/the-surprising-benefit-of-work-sabbaticals/#70a4432774d3

Butt, Mark, "The View from the Bridge," *Reservist*, Vol. 61, No. 1, 2014, p. 6.

Campion, Lisa, "Study Clarifies Job-Rotation Benefits," *Workforce*, November 1, 1996. As of April 28, 2020:
https://www.workforce.com/news/study-clarifies-job-rotation-benefits

"Career Development Program Requirements for Foreign Service Generalists," handout from U.S. Department of State, undated.

Carter, Ashton, Secretary of Defense, "Force of the Future: Maintaining Our Competitive Advantage in Human Capital," memorandum for secretaries of the military departments, November 18, 2015.

———, "Remarks on 'Goldwater-Nichols at 30: An Agenda for Updating,'" speech, Center for Strategic and International Studies, Washington, D.C., April 5, 2016.

Cartwright, Susan, and Cary L. Cooper, "The Role of Culture Compatibility in Successful Organizational Marriage," *Academy of Management Perspectives*, Vol. 7, No. 2, 1993, pp. 57–70.

Charman, A., *Global Mergers and Acquisitions: The Human Resource Challenge*, Alexandria, Va.: Society for Human Resource Management, 1999.

Clark, P. B., and J. Q. Wilson, "Incentive System: A Theory of Organization," *Administrative Science Quarterly*, Vol. 6, 1961, pp. 129–166.

Commander, Naval Reserve Forces (COMNAVRESFOR), "Command Brief," April 22, 2016.

COMNAVRESFOR—*See* Commander, Naval Reserve Forces.

Coss, Michael A., "Joint Professionals: Here Today, Here to Stay," *Joint Forces Quarterly*, No. 38, July 2005, pp. 92–99.

Crowe, William J., Jr., *The Line of Fire*, New York: Simon & Schuster, 1993.

Davenport, Christian, "Air Force Plan to Get Rid of A-10s Runs into Opposition," *Washington Post*, April 10, 2014. As of April 28, 2020:
https://www.washingtonpost.com/business/economy/air-force-plan-to-get-rid-of-a-10s-runs-into-opposition/2014/04/10/de0f041c-c015-11e3-b574-f87488/1856a_story.html

Defence Reserves Association, "Defence White Paper 2015," p. 11.

Defense Intelligence Agency, "DIA's Talent Management System Overview," briefing slides, Washington, D.C., 2016.

———, "FY17 Annual Career Assignment Program (ACAP) Way Forward," briefing slides, Washington, D.C.: DIA Office of Human Resources, April 8, 2016.

Deloitte, *Big Demands and High Expectations: The Deloitte Millennial Survey— Executive Summary*, New York, January 2014, p. 3. Accessed December 14, 2018: https://www2.deloitte.com/content/dam/Deloitte/global/Documents/About-Deloitte/gx-dttl-2014-millennial-survey-report.pdf

de Noble, Alex F., Loren T. Gustafson, and Michael Hergert, "Planning for Post-Merger Integration—Eight Lessons for Merger Success," *Long Range Planning*, Vol. 21, No. 4, August 1988, pp. 82–85.

Department of Defense Directive 1200.17, *Managing the Reserve Components as an Operational Force*, October 29, 2008.

DIA—*See* Defense Intelligence Agency.

DiFonzo, N., and P. Bordia, "A Tale of Two Corporations: Managing Uncertainty During Organizational Change," *Human Resource Management*, Vol. 37, 1998, pp. 295–303.

DoD—*See* Department of Defense.

Edmunds, Timothy, Antonia Dawes, Paul Higate, K. Neil Jenkings, and Rachel Woodward, "Reserve Forces and the Transformation of British Military Organisation: Soldiers, Citizens and Society," *Defence Studies*, Vol. 16, No. 2, 2016, pp. 118–136.

Eisenhower, Dwight D., "Address at the Centennial Celebration Banquet of the National Education Association," Washington, D.C., April 4, 1957.

Epstein, Marc, "The Drivers of Success in Post-Merger Integration," *Organizational Dynamics*, Vol. 33, No. 2, May 2004, pp. 176–179.

Fernandez, Sergio, and Hal G. Rainey, "Managing Successful Organizational Change in the Public Sector," *Public Administration Review*, March/April 2006, pp. 168–176.

Garone, Elizabeth, "The Surprising Benefits of a Mid-Career Break," *BBC*, March 28, 2016. As of April 28, 2020: https://www.bbc.com/worklife/article/20160325-the-surprising-benefits-of-a-mid-career-break

Goldrich, Robert L., *The Army's Roundout Concept After the Persian Gulf War*, Congressional Research Service Report for Congress, October 22, 1991.

Hagee, Michael W., Commandant of the Marine Corps, Statement to U.S. Senate Committee on Armed Services, *Defense Authorization Request for Fiscal Year 2006 and the Future Years Defense Program*, 109th Cong., 1st sess., February 10, 2005.

Holloway, J. L., III, "[Iran Hostage] Rescue Mission Report," August 1980.

Holt, Daniel T., Achilles A. Armenakis, Hubert S. Field, and Stanley G. Harris, "Readiness for Organizational Change: The Systemic Development of a Scale," *Journal of Applied Behavioral Science*, Vol. 43, 2007, pp. 232–255.

Intel, "Rotation Program." *Intel.* As of January 10, 2018: www.intel.com/content/www/us/en/jobs/student-center/rotations.html

James, Deborah, Secretary of the Air Force, "Memorandum of Instruction for Management Level Reviews," March 18, 2016.

Kansal, Sugandh, and Arti Chandani, "Effective Management of Change During Merger and Acquisition," *Procedia Economics and Finance*, Vol. 11, 2014, pp. 208–217.

Kanter, Rosabeth Moss, Barry A. Stein, and Todd D. Jick, *Challenge of Organizational Change: How Companies Experience It and Leaders Guide It*, New York: Free Press, 1992.

Kotter, John P., *A Force for Change: How Leadership Differs from Management*, New York: Free Press, 1990.

Lambright, W. Henry, "Leadership and Change at NASA: Sean O'Keefe as Administrator," *Public Administration Review*, March/April 2008, pp. 230–240.

Leguizamon, Stephanie, U.S. Marine Corps Forces Reserve, "Reserve Marines Prove Readiness to Support the Active Component at ITX 4-17," Twentynine Palms, Calif., July 3, 2017.

Lewis, L. K., and D. R. Seibold, "Reconceptualizing Organizational Change Implementation as a Communication Problem: A Review of Literature and Research Agenda," in M. E. Roloff, ed., *Communication Yearbook 21*, Beverly Hills, Calif.: Sage, 1998, pp. 93–151.

Lind, William S., "JCS Reform: Can Congress Take On a Tough One?," *Air University Review*, September–October 1985 pp. 47–50.

Locher, James R. III, "Taking Stock of Goldwater-Nichols," *Joint Force Quarterly*, Autumn 1996, pp. 10–11.

Lucas, Colleen, and Theresa Kline, "Understanding the Influence of Organizational Culture and Group Dynamics on Organizational Change and Learning," *The Learning Organization*, Vol. 15, No. 3, 2008, pp. 277–288.

Martin, Carol A., "From High Maintenance to High Productivity: What Managers Need to Know About Generation Y," *Industrial and Commercial Training*, Vol. 37, No. 1, 2005, pp. 39–44.

Milley, Mark A., Gen., Chief of Staff of the Army, "Winning Matters, Especially in a Complex World," Association of the United States Army, October 5, 2015. As of April 28, 2020: https://www.ausa.org/articles/winning-matters-especially-complex-world

Moran, John W., and Baird K. Brightman, "Leading Organizational Change," *Journal of Workplace Learning*, Vol. 12, No. 2, 2000, pp. 66–74.

National Association of Colleges and Employers, "Rotational Programs Yield Higher Retention Rates," March 22, 2017. As of April 28, 2020: https://www.naceweb.org/talent-acquisition/onboarding/rotational-programs-yield -higher-retention-rates/

National Commission on the Future of the Army, NCFA Operation Subcommittee Report, Open Meeting, *The Total Force Policy and Integration of Active and Reserve Units (Multiple Component Units-MCU)*, December 17, 2015.

———, *Report to the President and the Congress of the United States*, January 28, 2016.

National Commission on the Structure of the Air Force, *Report to the President and Congress of the United States*, Arlington, Va., January 30, 2014.

Navas, William A., Jr., "Integration of the Active and Reserve Navy: A Case for Transformational Change," *Naval Reserve Association News*, No. 5, 2004.

NCFA—*See* National Commission on the Future of the Army.

NCSAF—*See* National Commission on the Structure of the Air Force.

ODNI—*See* Office of the Director of National Intelligence.

Office of the Director of National Intelligence, "Joint Duty." As of January 10, 2018: https://www.icjointduty.gov/faq.htm

Office of the Secretary of Defense for Policy, homepage. As of January 10, 2018: http://policy.defense.gov

Oreg, Shaul, "Personality, Context and Resistance to Organizational Change," *European Journal of Work and Organizational Psychology*, Vol. 15, No. 1, 2006, pp. 73–101.

O'Shaughnessy, John, *Patterns of Business Organization*, London: Routledge, 2013.

Pellerin, Cheryl, "Carter Unveils Next Wave of Force of the Future Initiatives," *DoD News*, June 9, 2016. As of April 28, 2020: https://www.defense.gov/Explore/News/Article/Article/795625/carter-unveils-next-wave-of -force-of-the-future-initiatives/

Peterson, Eric, 120th Airlift Wing Public Affairs Office, "Restructuring Brings New Capabilities to the 219th RED HORSE Squadron," June 2, 2017. As of January 11, 2018: http://www.120thairliftwing.ang.af.mil/News/Article-Display/Article/1201591/ restructuring-brings-new-capabilities-to-the-219th-red-horse-squadron/

Phillips, Mark, *The Future of the UK's Reserve Forces*, London: RUSI, April 2012.

Pink, D. H., *Drive: The Surprising Truth About What Motivates Us*, New York: Riverhead Books, 2009.

Public Law 99-433, Goldwater-Nichols Department of Defense Reorganization Act of 1986, October 4, 1986.

Quast, Lisa, "Overcome the 5 Main Reasons People Resist Change," *Forbes*, November 26, 2012. As of April 28, 2020:
https://www.forbes.com/sites/lisaquast/2012/11/26/overcome-the-5-main-reasons-people-resist-change/#4d44ad6b3efd

Reserve Forces Policy Board, *Reserve Component Use, Balance, Cost and Savings: A Response to Questions from the Secretary of Defense: Final Report to the Secretary of Defense*, RFPB Report FY14-02, Falls Church, Va., February 11, 2014.

RFPB—*See* Reserve Forces Policy Board.

Rhodes, Phillip F., "Reservists Selected to Lead Active-Duty Units," *Air Reserve Personnel Center*, December 23, 2015. As of January 10, 2018:
http://www.arpc.afrc.af.mil/News/ArticleDisplay/tabid/267/Article/637785/reservists-selected-to-lead-active-duty-units.aspx

Robbert, Albert A., James H. Bigelow, John E. Boon, Jr., Lisa M. Harrington, Michael McGee, S. Craig Moore, Daniel M. Norton, and William W. Taylor, *Suitability of Missions for the Air Force Reserve Components*, Santa Monica, Calif.: RAND Corporation, RR-429-AF, 2014. As of December 18, 2018:
https://www.rand.org/pubs/research_reports/RR429.html

Robbert, Albert A., Tara L. Terry, Paul D. Emslie, and Michael Robbins, *Promotion Benchmarks for Senior Officers with Joint and Acquisition Service*, Santa Monica, Calif.: RAND Corporation, RR-1447-OSD, 2016. As of December 18, 2018:
https://www.rand.org/pubs/research_reports/RR1447.html

Rogelberg, Steven G., *The SAGE Encyclopedia of Industrial and Organizational Psychology*, Thousand Oaks, Calif.: Sage Publications, 2007.

Ruhling, David, "Shaping the Reserve Workforce," *Reservist*, Vol. 61, No. 1, 2014, p. 24.

Schuler, Randall, and Susan Jackson, "HR Issues and Activities in Mergers and Acquisitions," *European Management Journal*, Vol. 19, No. 3, 2001, pp. 239–253.

Schweiger, D. M., and A. S. Denisi, "Communication with Employees Following a Merger: A Longitudinal Field Experiment," *Academy of Management Journal*, Vol. 34, No. 1, 1991, pp. 110–135.

Secretary of the Air Force Public Affairs, "Air Force Continues to Pursue Total Force Integration," *U.S. Air Force*, March 11, 2016. As of January 10, 2018:
http://www.af.mil/News/Article-Display/Article/691270/air-force-continues-to-pursue-total-force-integration/

"Send in the Reserves," *Armed Forces Journal*, February 1, 2012. As of April 28, 2020:
http://armedforcesjournal.com/send-in-the-reserves/

Senge, Peter M., *The Fifth Discipline: The Art and Practice of the Learning Organization*, New York: Doubleday, 1990.

Serbu, Jared, "Army to Experiment with New Blended Units of Active, Reserve Forces," *Federal News Radio*, March 22, 2016.

Thompson, Loren, "Shrinking Army Fights National Guard for Vital Combat Helicopters," *Forbes*, June 30, 2014. As of April 28, 2020:
https://www.forbes.com/sites/lorenthompson/2014/06/30/shrinking-army-fights-national-guard-for-vital-combat-helicopters/#3cc97412566e

UK Ministry of Defence, *The Independent Commission to Review the United Kingdom's Reserve Forces*, London, July 2011.

———, "Consultation Launched on the Future of Britain's Reserve Forces," November 8, 2012. As of January 10, 2018:
https://www.gov.uk/government/news/consultation-launched-on-the-future-of-britains-reserve-forces

U.S. Coast Guard, "Coast Guard Reserve History," December 7, 2018. As of April 28, 2020:
https://www.reserve.uscg.mil/about/history/

———, Commandant Instruction 5320.4A, *Reserve Force Readiness System (RFRS) Staff Element Responsibilities*, November 6, 2014.

U.S. Department of Air Force, Air Force Instruction 90-1001, *Special Management: Planning Total Force Associations*, January 9, 2017.

———, Air Force Policy Directive (AFPD) 90-10, *Total Force Integration Policy*, June 16, 2006 (certified current July 31, 2014).

———, Air Force Policy Memorandum (AFPM) 90-10, *Total Force Integration*, October 27, 2016.

U.S. Department of Air Force, Secretary of Air Force Public Affairs, "AF Announces Stand Up of Integrated Wing," Washington, D.C., February 10, 2016.

U.S. Department of Army, Army Directive 2012-08, *Army Total Force Policy*, September 4, 2012.

U.S. Department of Defense, "Force of the Future: Whatever You Want to Do, You Can Do in Service to Your Country." As of January 11, 2018:
https://www.defense.gov/News/Special-Reports/0315_Force-of-the-Future/

U.S. Government Accountability Office, *Defense Management: Fully Developed Management Framework Needed to Guide Air Force Future Total Force Efforts*, GAO-06-232, Washington, D.C., January 31, 2006.

U.S. Marine Corps Forces Reserve, "Definitions." As of January 16, 2018:
http://www.marforres.marines.mil/Major-Subordinate-Commands/Force-
Headquarters-Group/Marine-Corps-Individual-Reserve-Support-Activity/
Definitions/

Weitz, Richard, *The Reserve Policies of Nations: A Comparative Analysis*, Carlisle,
Pa.: Strategic Studies Institute, September 2007.

Wilson, J. Q., *Bureaucracy: What Government Agencies Do and Why They Do It*,
New York: Basic Books, 1989.

Winkler, John D., et al., "A 'Continuum of Service' for the All-Volunteer Force,"
in Barbara A. Bicksler, Curtis L. Gilroy, and John T. Warner, eds., *The All-
Volunteer Force: Thirty Years of Service,* Washington, D.C.: Brassey's, Inc., 2004,
pp. 297–307.